The Secret Lives of Numbers

For Susanna and Freddie,
both of whom can count better than me.

The Secret Lives of Numbers

The Curious Truth Behind Everyday Digits

Michael Millar

Illustrated by Louise Morgan

Virgin BOOKS

2 4 6 8 10 9 7 5 3 1

First published in 2012 by Virgin Books, an imprint of Ebury Publishing

A Random House Group Company

www.randomhouse.co.uk

Addresses for companies within The Random House Group Limited can
be found at www.randomhouse.co.uk/offices.htm

The Random House Group Limited Reg. No. 954009

A CIP catalogue record for this book is available from the British Library

The Random House Group Limited supports The Forest Stewardship Council
(FSC®), the leading international forest certification organisation.
Our books carrying the FSC label are printed on FSC® certified paper.
FSC is the only forest certification scheme endorsed by the leading
environmental organisations, including Greenpeace.
Our paper procurement policy can be found at:
www.randomhouse.co.uk/environment

Printed and bound by CPI Group (UK) Ltd, Croydon, CR0 4YY

ISBN: 9780753540862

To buy books by your favourite authors and register for offers, visit:
www.randomhouse.co.uk

CONTENTS

INTRODUCTION

First things first: this book is not boring. Yes, it's a book about numbers and, for many people, that is a concept best resigned to the days of the schoolroom, ne'er to be seen again. But stay with me. In fact, this book is far from dull; instead, it will make you see various parts of your daily life in a whole new light.

This is a book about the past, the present, and the future. Indeed, it goes right up until the end of the world . . . It is about culture, sport, music, work – and bouts of heavy drinking.

After numerous failed attempts to sum up for friends exactly what I was writing, I settled on the straightforward moniker of 'a history book about numbers'. In some senses that is quite apt; the origins of many of the numbers featured in these pages stretch back a long, long way – sometimes millennia.

But I think calling it a 'history book' is also a disservice – because it is about much more than just the happenings in days of yore. It is about things that shape your modern, twenty-first-century life on a daily basis. Certain habits and numbers that we accept as normal – mundane, even – might have been adopted at the beginning of recorded time, but that doesn't stop them having a direct impact on questions like 'Why are you on your way to work today?' And, indeed, 'Why is today, today at all?'

I've chosen the numbers featured in this book based on the fact that pretty much all of them are likely to impinge on you in one way or another during your life. Of course, some will have more direct impact than others; practising Satanists might find themselves drawn to '666' before they turn to the chapter on tennis scores.

These entries have to include a certain amount of informed guesswork. After all, many of the customs and numbers that form integral parts of our lives – such as the seven-day week – were set in place so long ago that we're really not sure exactly why they are there. You can criticise this unfortunate truth all you like, but until you bring me an example of ancient man mulling over how many fingers and toes he might or might not have, then we'll have to agree to disagree.

Nonetheless, while some numbers are shrouded in mystery, others charge out at you as clear as day. They are the numbers on the back of your football shirt; they are the credit cards that keep us in the manner to which we are accustomed; they are the Sudoku puzzles that keep us entertained on the way to and from work. This book will hopefully make them more than just random items in your day-to-day life.

If all else fails, the information inside is sure to help in a pub quiz at some time or another.

Michael Millar

9

BEHIND CLOUD NINE

In case you haven't been fortunate enough to go there, Cloud Nine is a metaphorical place of great joy or euphoria. The origin of this expression is often attributed to a system of cloud numbering created in the 1950s by the US Weather Bureau. However, while it is eminently possible that meteorology was behind Cloud Nine, to attribute the system to the US Weather Bureau is incorrect, as the original classification actually came in the *International Cloud Atlas* of 1896.

High as a Kite

The *Atlas* classification gave the cumulonimbus cloud the number nine out of a list of ten. This cloud is one of the highest (reaching heights of 45,000 feet) and is certainly the fluffiest and most comfortable-looking of the clouds.

Cloud Nine therefore seemed a nice metaphor for floating, carefree, high above the Earth (and was enthusiastically adopted by marijuana smokers during the 1960s).

> We should point out that the tenth cloud in the list is the 'stratus', which is the lowest-forming of all the cloud types and is basically just boring, featureless mist. This in itself is a useful metaphor to show how easily you can fall back to Earth if you're not careful!

Many people attribute the original popularity of the phrase to a 1950s radio show called *Johnny Dollar*, in which the hero was transported 'to Cloud Nine' whenever he was knocked out.

The Cloud Next-Door-But-One

Despite the current popularity of the expression Cloud Nine, there is also evidence that Cloud Seven was once a popular term for describing happiness. It may well have been the cloud of choice for joyful folk long before Cloud Nine got so popular it had to start putting up 'no vacancy' signs. For example, the 1960 edition of the *Dictionary of American Slang* described living on Cloud Seven as being 'completely happy, perfectly satisfied, in a euphoric state'.

This may have its origins in 'seventh heaven', another term for great joy. This in turn came from the belief in several religions that the seventh is the highest of all the realms of heaven, where God and the angels live.

Visiting Cloud Eight to Get to Cloud Nine

Finally there is another cloud, which really doesn't get the recognition it deserves: Cloud Eight. In fact you may well have visited it yourself without realising it. In the 1930s, if you were said to have been on Cloud Eight it meant that you had had too much to drink.

The Underworld Speaks, a book published by the FBI in 1935 to help police spot gangsters by how they talked, defined the expression Cloud Eight as 'befuddled on account of drinking too much liquor'.

This might explain why hangovers hurt so much: you've just plummeted off a cloud, after all.

5

THE FIVE-STAR
HOLIDAY MYTH

The plaque proclaiming five-star status is likely to be emblazoned with pride in the foyer of the luxury hotel you have chosen for your holiday (or perhaps in the lobby of the building you look at longingly as you wander back to Pedro's Bargain Hotel & Sewage Works).

Tourists tend to believe such ratings offer an accredited assessment of the establishment they have chosen, providing them with reassurance about what to expect upon arrival. Little do most travellers know: the stars that guide them could be seriously misleading.

Putting Your Faith in the Stars

The premise is simple: the higher the standards of cleanliness, ambience, hospitality, service and food at the hotel, the more stars an establishment earns. But the wisdom of

4

putting your faith in the star rating of a hotel will depend on where in the world you are travelling.

In fact, there's just as much chance that your hotel will be graded by letters or diamonds as stars. This is because, contrary to widespread belief, there is no international standard to rate hotels – and in some cases the hotels award the accolades to themselves.

What qualifies a hotel for a certain number of stars (or diamonds or numbers for that matter) varies by global region, country, and in many cases what area of a country you are in.

> **There are also questions around just how high the ratings can go. Until recently five-star status was a byword for top-end luxury, but a growing number of hotels boast six or even seven stars. In fact there is talk of a hotel in the Middle East that will lay claim to ten.**

From Brandy to Bedrooms

Probably the oldest hotel rating system in the world was introduced in the United Kingdom in 1912, when the Automobile Association's secretary, Stenson Cooke, came up with the idea.

The official story goes that he once worked as a wine and spirit salesman and 'felt that the star rating of brandy would be a familiar yardstick to apply to hotels'. Thus a three-star classification system was born, which evolved into a five-star system now overseen jointly by the AA and the national tourist bodies of each area of the UK. However, even with a major head start over successors around the world, this system was not harmonised until 2007.

Diamonds or Stars?

The United States has several competing awards, including those of the American Automobile Association, which assigns diamonds, and Forbes Travel, which assigns stars.

Forbes send in inspectors who judge the hotel based on a range of classifications, such as graciousness, efficiency and luxury. For every negative, a hotel loses points, and they're pretty tough in their assessment.

So if you go to a Forbes-rated hotel, you're likely to know what you're getting. But to put this in context, there are approximately 50,000 hotels in operation today in North America, whereas *Forbes Travel Guide* only recommends around 3,000. So in many cases you will be on your own. In fact there's nothing to stop you rating the other 47,000 or so yourself if you have some time on your hands.

Something for Everyone

The story of competing systems, contrasting systems and systems that are just plain made up is one that is repeated across the globe. If you book a room in Turkey, for example, you might face a choice between the rating system run by central government and one controlled by local municipalities.

One of the most coherent cross-border approaches is found in Europe, where 11 countries have banded together to create the Hotelstars Union classification system. This is based on a traditional star system: Tourist (*), Standard (**), Comfort (***), First Class (****) and Luxury (*****).

The system involves 21 qualifications encompassing 270 elements, where some are mandatory for a star and others optional.

To get one star the hotel must have the following:

- A shower/WC or bath tub/WC in each room
- Colour TV and remote control, and a table and chair in each room
- Soap or body wash provided
- Daily room cleaning
- Reception service
- Public telephone and facsimile for guests
- Extended breakfast available
- Beverages on offer within the hotel
- The option to leave items securely at reception

As the ratings build, everything from a reading light next to the bed and bath towels (among the requirements for two stars) to a bathrobe and slippers on demand (for four stars) is included. However, for the most discerning among you, the key things that separate a five-star hotel from a four-star hotel are:

- Reception open 24 hours, with multilingual staff
- Doorman service or valet parking
- A concierge
- A spacious reception hall with several seats and a beverage service
- A personalised greeting for each guest with fresh flowers or a present in the room

- A minibar in the room and food and beverage offered via 24-hour room service
- Personal care products in 'flacons' (those small stoppered bottles)
- Internet connection in the room
- A safe in the room
- An ironing service (returned within one hour) and shoe polish service
- Turndown service in the evening

The Sky's the Limit

This brings us to the extraordinarily opulent hotels that claim, or have been attributed, six- and even seven-star status. No formal body awards or recognises any rating over 'five-star deluxe', so be wary of such claims as they are predominantly used for advertising purposes.

Nonetheless, it would be a mistake to think that these establishments don't try to live up to the name. The Emirates Palace in Abu Dhabi uses 5 kg of edible gold each year to decorate food. Australia's Palazzo Versace is not only packed with the designer's furniture and fittings, it has its own 90-berth marina, and the Indonesian resort of Amanjiwo offers you a live-in butler.

Quite simply, there is something for everyone wherever you travel. Just make sure you look at more than the star rating to be sure you get what you think you are paying for. The moral of the story: following stars doesn't necessarily make you a Wise Man.

2-3-5

WHY THE GOALKEEPER WEARS 1

*I*n 1911 there was a sensational innovation at a local "Australian Football" match in Sydney,' records the International Federation of Football History & Statistics.

What was this extraordinary change that set the footballing world agog in 1911? Shorts that daringly exposed the players' knees? Perhaps a new ball that gave you a fighting chance of not breaking your neck when you headed it?

No, it was something that is now present as a matter of course in every game: shirt numbers.

Making Football History

The Federation reports that Sydney soccer clubs Leichhardt and HMS *Powerful* were the first official teams to follow suit, wearing squad numbers on their backs in the very same year, and making football history as they did so.

Numbers were quite simply forms of identification so the fans and journalists present at the games could work out who was doing what.

Behind-the-Times Brits

This event seems to pour cold water on English claims that shirt numbers were their innovation. In fact, it wasn't until seventeen years later that English teams got in on the act, when, in August 1928, The Wednesday (renamed Sheffield Wednesday shortly after) took on Arsenal and Chelsea played Swansea Town.

Both teams sported numbers on their shirts, marking the earliest recorded use of shirt numbers in Football League play.

The Shirt Number System

Five years later, on 29 April 1933, shirt numbers were worn for the first time in the Football Association Challenge Cup final. But the system was not quite as we know it now: Everton players wore numbers 1 to 11, while Manchester City had numbers 12 through 22.

In 1939 the Football League decided all teams should wear numbers 1 to 11, with each number pertaining to a particular position.

However, this decree was limited in its initial impact as the football season was almost immediately interrupted by the small matter of the Second World War.

> **Some clubs reserve a number for their fans, refusing to issue it to any player. The most common number used is 12, since the fans in a stadium are often given the moniker of 'The Twelfth Man' due to the boost their support gives the team.**

What's Your Number?

There is some confusion about how the shirt numbering system worked, as the number 5 is often found to be in the centre of defence, for what appears to be no reason whatsoever.

As is so often the case, the reason for this anomaly is rooted in numerical evolution. Until the 1960s, football teams regularly played in a 2-3-5 formation. When this strategy was laid out in match programmes or in pre-match newspaper reports, the goalkeeper, at number 1, was at the top, giving it a distinctly Christmas tree-esque appearance.

The numbers then ran from left to right as you went down the 'tree', until they reached 11 on the left wing.

Of course, formations changed radically as time went on; one change being that the centre half, who was number 5, dropped back to centre back. This put him (it invariably was a 'him' back then) between the other two backs, numbers 2 and 3. And thus the confusion was born.

Squad Shirts

It wasn't long until teams, for organisational and marketing purposes, started assigning squad numbers instead of just starting every game with numbers 1 to 11, regardless of who was playing. In 1954, FIFA's World Cup Final competition regulations required teams taking part in the tournaments to adopt a squad numbering system.

A 1965 change in the rules, meanwhile, brought even more numbers onto the pitch, as substitutions were allowed for the first time. To start with, this was only to replace injured players, but the next year this was extended to allow them for any reason.

> **Along the way football teams have tended to mess about with numbering in any way that they see fit. For example, in 1974 the Dutch national team became one of a number of sides to hand out numbers in alphabetical order. This led to striker Ruud Geels getting the number 1 shirt, which is usually reserved for goalkeepers.**

Name and Number

It took a remarkably long time for the marketing men to cotton onto the fact that they could make more money by adding players' names to shirts as well. This didn't happen until the Premier League was introduced in 1992.

It could be argued that football numbers have now been superseded in importance by the names of the superstar players that sit above them.

29

THE LEAP DAY

Are you looking to point the finger of blame at someone for making you work an extra day every four years, almost certainly without extra pay? Well, you can start by pinning the leap year on our solar system's poor time-keeping.

Everyone knows that a year – the amount of time it takes the Earth to complete one orbit of the Sun – is 365 days. But, as everyone also knows, this isn't quite right because this orbit does not add up to an exact number of whole days. Rather, our annual solar tour takes 365.2422 days. This means every four years a leap year is added to the Gregorian calendar (which most of the world uses) to compensate for these extra hours.

13 Months a Year?

But it hasn't always been this way. Back in Julius Caesar's day there was a 355-day calendar, which had an extra 22-

day month added every two years. However, this was complicated to administer, particularly because important festivals were sometimes pushed into the wrong seasons. (There is also evidence that the priests in charge of the calendar meddled with it to keep friendly officials in power for longer, while giving enemies shorter tenure.)

Caesar, who clearly didn't want his roaring, gorging and orgy activities beset by such confusion, demanded that his astronomer, Sosigenes of Alexandria, should sort this out. The star-gazer decided on a 365-day year with an extra day every four years to cover the additional hours. This 'Julian calendar' kicked in at the start of 45 BC.

But there was a problem. If the extra hours for which the calendar was compensating were exactly a quarter of a day per year, then adding an extra day every four years would regulate the calendar perfectly. But they aren't, they're slightly short (by a pesky 0.0078 of a day per year – that's just over 11 minutes).

The Gregorian Solution

It took over 1500 years for this to be ironed out – not until Pope Gregory XIII and his astronomers introduced the Gregorian calendar in 1582. The solution they lit upon was to lose three leap days every 400 years.

The rule is that if a year is divisible by 100, but not by 400, it is not a leap year. Following this formula we see the dates 1700, 1800 and 1900 were not leap years, whereas 2000 and 1600 were. Scholars say this impressive feat of maths will work for another 10,000 years or so, when a rethink will be needed.

The Lost 10 Days

When the new calendar came in, the year suddenly leaped forward by 10 days in October 1582, causing some to think a week and a half of their life had just been taken away. They were not impressed.

Others simply refused to play ball; Britain stuck with the Julian calendar until 1751, at which time the New Year started on 25 March, known as 'Lady Day'. This meant, for example, the day after 24 March 1712 was 25 March 1713. Confusing.

A Leap of Faith

The term 'leap year' is one that defies a comprehensive explanation. One answer is it originated in fourteenth-century Britain where the extra day caused fixed festivals, which normally advanced one weekday per year, to 'leap' ahead one day in the week. However, this is far from certain.

What most people don't ask is why it is 29 February that is the chosen 'extra' day. Again this was down to a Roman emperor sticking his aquiline nose into things.

When Caesar Augustus came to power, February had 29 days and the 30th was set aside for leap years. But Augustus was upset that the month named after him (August, obviously) only had 30 days, rather than the 31 boasted by his predecessor, Julius Caesar's, July. So February had a day removed since it was already out of whack with the alternating 30/31-day system that the Julian calendar had created.

But this meant that Romans were stuck with three months in a row that had 31 days. So officials switched the lengths of the last four months around, so now '30 days have September, April, June and November', as per the rhyme.

> **Other emperors attempted to achieve immortality through the calendar with less success. If they had, we would still have 'Claudius' instead of May and 'Neronius' instead of April.**

Lock Up Your Sons!

The 'leap day' is traditionally held to be the one day of the year when it is acceptable for women to propose to men. It is not clear where this tradition came from but one of several stories revolves around St Bridget complaining to St Patrick in the fifth century that men were taking too long to pop the question.

According to this account, he had sympathy with her but clearly wanted to limit the impact on his fellow man, thus decreeing that 29 February should be the day that women were allowed to propose.

Another story comes from thirteenth-century Scotland, where Queen Margaret was said to have brought in a law setting fines (which included a kiss, strangely enough) for men who turned down marriage proposals made by women on a leap day.

Since she was resident in Norway and only five years old at the time this was said to have happened, it seems unlikely that Margaret was so aware of this problem faced by the sisterhood that she felt it necessary to change the law. Also, the fine of a kiss seems pretty inappropriate in the circumstances.

A third theory proposes that once upon a time English law did not recognise the concept of a leap year, meaning that women could use this legal loophole to propose.

Probably the best place for a woman to take advantage of the rule is in Denmark, where the tradition actually falls on 24 February each leap year. If she is turned down, custom dictates the man gives her twelve pairs of gloves – one for each month to cover up the lack of ring on her finger.

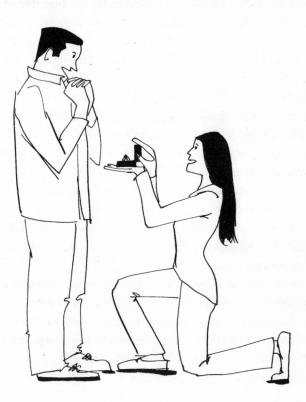

13

UNLUCKY FOR SOME

If you are worried about the number 13 being unlucky, then it can make life very difficult. In fact, you'd probably be best off living in the elevator of a United States hotel if at all possible, since they very often don't have a button for the 13th floor, jumping instead from 12 to 14. (Whether there actually is a secret 13th floor that can only be reached by the stairs is something no one seems inclined to ask.)

There is also the hassle of explaining your problem to others, as the fear of number 13 is technically known as 'triskaidekaphobia', which is quite a mouthful.

Such concerns mean you may have to make numerous adjustments in daily life. For example, some people believe that to have 13 people around a dinner table is unlucky, and tradition dictates that a teddy bear or the like be given a seat to increase the numbers. Superstition suggests that refusing teddy his chair means that one of the guests will die – which

leads us to the first of many explanations for 13 having such a bad reputation.

Deadly Dinner Dates

There are two dinner invitations that have gone down in history as being to blame for 'unlucky number 13'. The first is the Last Supper, where the Bible tells us that Jesus and his 12 disciples (making a total of 13, obviously) met for the last time to eat before he was betrayed and then crucified.

The other infamous supper date is from Norse mythology, where the most mischievous of the gods, Loki, arrived uninvited to dinner, becoming the 13th guest. During the course of the evening, he tricked one god into killing another. The latter happened to be associated with beauty, joy and righteousness, making the freeloading Loki very unpopular indeed.

Reasons to be Fearful

If you're not a religious person, of course, fear not – or rather do: there are still plenty of reasons why you can be scared of the number 13.

It was said that a hangman worth his salt would ensure that his noose had at least 13 turns of the rope to be certain of breaking a neck, while witches were believed to gather in covens of 13, of which the 13th member could be the Devil.

Some say the fear is rooted deep in our subconscious, as early man could count no higher than the ten fingers on his hands and his two feet, meaning his numeric abilities stopped at 12 – anything beyond that was a mystery. Another mystery is why he would ignore his toes.

Having 13 letters in your name is said to trip you up, with advocates citing killers from Jack the Ripper to Charles Manson. What Nelson Mandela thinks of this theory is unclear.

Perhaps the most likely reason for 13 having a bad name, however, is quite simply because it is the poor neighbour of number 12. From ancient cultures to the modern day, 12 has been seen as a number of completion. There are 12 hours on the clock face, 12 months in a year and 12 signs of the zodiac. This being the case, 13 ventures into the unknown and messes about with the space–time continuum.

> On 13 October 1307 (incidentally a Friday), the King of France ordered the arrest of the Knights Templar, many of whom were horrifically tortured or killed. This event is often put forward as the origin of 'Friday the 13th' being particularly bad luck.

Lucky 13

But the number 13 does not suffer from such a scurrilous reputation everywhere. In China, for example, it is among the luckiest of numbers, and Chinese people often pay huge amounts for items that contain it, such as vehicle number plates.

This is down to the fact that 13, like several lucky numbers in China, is pronounced in a similar way to other words that have positive connotations. When the numbers 1 and 3 are combined to make 13, the resulting word – which in Mandarin is pronounced *'shisan'* (*shi* and *san* combined) – could also mean 'definitely vibrant' or 'assured growth'.

Instead the Chinese look on number 4 as particularly unlucky, because it sounds very like the word for 'death'.

The Number of Eternal Life

The ancient Egyptians were also said to be big fans of the number 13. They believed life was made up of 12 stages, before they reached the 13th – eternal life. So while 13 symbolised death in one sense, it also meant rebirth in another.

Some people in Western culture have made a point of laughing in the face of this superstition, athletes in particular. Legendary American football quarterback Dan Marino wore 13 on his shirt, while basketball superstar Wilt Chamberlain put his own spin on it, wearing number 13 to show how unlucky his opponents were.

Business (Mis)Fortunes

In 2004, the Stress Management Center and Phobia Institute in North Carolina, USA, claimed that between $800m and $900m was lost to industry each year because of people either being unwilling to fly on Friday 13th – or just refusing to get out of bed, full stop.

The Center estimates that 17–21 million Americans could be affected in one way or another, so any cynics trying to do business on that day may well find the number impacts on their success, whether they like it or not.

A Final Word of Warning

Some sceptics might say that this mixture of history, myth and coincidence means they can turn their nose up at unlucky 13 without any fear of the consequences. But they could be wrong, particularly if they go out unprepared on Friday 13th.

A study published in the British Medical Journal in the early 1990s showed there had been significantly less traffic on the UK's busiest motorway – the M25 – on a Friday 13th compared with the Friday before. Despite this, the risk of being hospitalised on that day due to a transport accident increased 52 per cent.

Coincidence? Perhaps . . .

52

COUNTING CARDS

There has no doubt been some form of cards played for as long as there's been money to lose. But the International Playing-Card Society reckons the earliest decks must have come from China, since they invented paper (you can't fault that logic).

Early Chinese cards had suits, which comprised different coins, with further coins to denote their place within that suit. The decks migrated westward and were first evident in Europe in the late fourteenth century.

They arrived courtesy of the Mameluks of Egypt, by which time they already consisted of 52 cards of four suits, though not ones we would recognise today – the Egyptian suits were swords, polo-sticks, cups and coins. The cards included numerals from one to ten and three 'Court Cards' – King, Deputy-King and Under-Deputy – reinforcing the status of the aristocracy at the top of the pile.

The European Card War

As joyful medieval Europeans succumbed to cards quicker than the plague, different variations of decks soon appeared, usually varying between 48 and 52 cards. Spanish suits were cups, swords, money and clubs, while the Germans favoured hearts, acorns, bells and leaves.

But it was the French design of hearts, spades, diamonds and clubs that became the most widely used and is the pattern most of us recognise today. This happened because the designs were simple to reproduce and they only had red and black suits, whereas other makers gave each suit a different colour. This meant the French could manufacture them cheaply.

Paving Stones and Pig Food

It is said the French suits reflected medieval society:

- The clergy was represented by hearts.
- The merchants got diamonds (because of the shape of the paving in the parts of churches where the rich were buried, rather than because of precious stones).
- The nobility were represented by spades, which were either based on knights' spearheads or hawk bells, dependent on whose story you believe.
- At the bottom of the pile were the peasants, who were represented by clubs, which evolved from clover: pig food.

Yet as nice a story as this is, there sadly doesn't seem to be any solid historical evidence for it.

Even the very earliest commercial decks of cards feature the numbers in the top corners of the cards. Why? Because this design ensures that players can keep their cards held in a tight fan shape and limit the scope for others players to spy on them.

Face Facts

One thing the early card manufacturers didn't do was create the 'double-ended' design we are familiar with now. It wasn't until the Victorian era that they did away with full-length cards (so, for example, until the nineteenth century, the Jack's head would have been at the top of the card and his feet at the bottom).

Otherwise, in terms of the 'face' cards, you'd be amazed how little they've changed in the last few hundred years. For example, cards from the late 1700s show kings looking regally off to the left or right in much the same way they do today.

What's in a Name?

Early card makers also assigned names to the face cards, which varied depending on who made them. In terms of who the kings were said to represent, there seems to have been a level of consistency from the early seventeenth century onwards among the French designers (many of whom were based in the same town, Rouen):

- David (spades)
- Charlemagne (hearts)
- Caesar (diamonds)
- Alexander the Great (clubs)

Jacks – or knaves as they were better known – were also given names that stuck:

- Ogier, of Arthurian legend (spades)
- La Hire, a heroic French knight (hearts)
- Hector of Troy (diamonds)
- Lancelot, again of Arthurian stories (clubs)

We won't go into who the queens were supposed to be because there is so much speculation, and unless you're a medieval historian there's a good chance you'll only ever have heard of Joan of Arc. (Empress Judith? Agnes Sorel? Isobel of Bavaria? No, thought not.)

In any case, all that discussion became slightly moot when the aristocracy fell out of fashion during the French Revolution of 1789. Since then, no manufacturers have officially applied names, though in a recent repeat of the tradition, the American military created a named deck during the last Iraq war. Each card from this notorious 'Most Wanted' pack represented a different member of Saddam Hussein's regime that they were trying to apprehend.

The Ace of Spades

The French Revolution is held to be one of the reasons the Ace rocketed from its lowly position as number one to the most powerful card in the pack. It is said that the Ace's promotion represented the common man overcoming the aristocracy.

This may be true, but the more likely sounding explanation is a bit more prosaic. In eighteenth-century England, you had to pay tax on cards; to prove manufacturers had paid up, officials stamped each deck – usually on the Ace of Spades. In 1765, the Stamp Office created an official Ace of Spades stamp, which incorporated the royal coat of arms. From then on, the Ace of Spades became the most ornate card in the pack – and its matinee-idol good looks landed it the leading role in many a card game.

Poker Scores: The Dead Man's Hand

Poker, of course, is a game where the importance of numbers is paramount, and many thousands of words have been written about how to calculate the odds to your advantage and which hands are worth betting on. But one combination of cards in particular has become infamous, with some believing it to be a harbinger of death: the Dead Man's Hand.

Famous gunfighter Wild Bill Hickok was playing poker in Deadwood, Dakota Territory, when he was shot from behind by Jack McCall on 2 August 1876. The legend says that the hand he was holding when he died at the Number 10 Saloon included two black aces and two black eights. While there is plenty of controversy over whether the story is true, the local barber who acted as undertaker for Wild Bill that day did later claim that's what he saw on the poker table.

The fifth card in Hickok's hand remains something of a mystery.

The barber/undertaker, one Ellis T. 'Doc' Peirce, reported all the other bullets in the assassin's gun were dud. If he had not been lucky enough to have chosen that particular chamber of his weapon, then he would probably have joined our famous numbers as the 37th man to be shot dead by Hickok. (Although the exact number of men Wild Bill killed is impossible to verify, and so this figure comes with a health warning, which seems appropriate in the circumstances.)

9.99

THE SECRET OF THE MISSING PENNY

You might think you're a savvy shopper and you'd see past age-old tactics like pricing something £9.99 instead of £10 to make it look better value. But you might be surprised by how susceptible to this little trick you actually are – and perhaps not in the way you expect.

This phenomenon is evident all over the world and for all our modern retail nous it still seems to work. As recently as 2008, academics in France studied what happened when a pizza was sold for €8.00 and €7.99. It turned out that knocking off that one cent increased sales by 15 per cent.

These results are all the more incredible in a world where inflation has made the extra penny you get back pretty useless – surely everyone has a pile of these horrible little coins lurking around their house somewhere, don't they?

The 99 Effect

There is a theory that this pricing system – often called the '99 effect' – began as an effort to combat employee theft. The story goes that costing something this way would require staff to open the till to get change, making it more difficult for them to pocket any bills handed over by the shopper.

But in studies into the origins of the '99 effect', Professor Robert Schindler, an academic at Rutgers Business School in the United States, has found the theft theory didn't seem to stand up.

He looked at adverts in the *New York Times* from 1850 and couldn't find any instances of prices ending in 99 cents. However, he did notice them sneaking in from the 1870s onwards, but only when shops were advertising discounts. The regular price tags remained round numbers.

Professor Schindler believes that somewhere along the line someone made the link between the spare penny and the perception of value, and this now ubiquitous trick was born.

> **Over a hundred years after the practice seemingly started in US department stores, a study in New Zealand in the late 1990s found around 60 per cent of prices in advertising material ended with a '9', while another 30 per cent ended with a '5'. Only 7 per cent ended with a nice round '0'.**

Tricks of the Trade

Knocking off the penny to make a price tag more attractive is known in the trade as 'psychological pricing'. The most straightforward attraction to the shopper is obvious: it looks slightly cheaper at a glance, giving the purchase an emotional incentive. (As such it is particularly powerful in terms of small or impulse purchases.)

This is particularly true in tough times when you will regularly see petrol prices, for example, cut by a penny to attract consumers, even though the difference this makes is negligible.

Knowing the Price of Everything and the Value of Nothing

There is another theory worth considering as to why we might be more willing to open our wallets and purses when retailers cut a penny off their prices. It revolves around how the brain thinks about value.

Researchers at the University of Florida did some tests and concluded that the more precise the price in terms of dollars and cents, the more shoppers could be persuaded to part with for an item.

For example, a kettle priced at £10, the theory goes, makes people think in round numbers: thus they would consider its price in terms of £10, versus £11 or £9. However, introducing pennies into the equation makes us think in terms of them instead; suddenly a £9.99 kettle becomes better value if it's £9.50. This means in practical terms that retailers have to knock a lot less off if they want to entice us.

The Florida academics tested this theory in the local housing market and found that sellers got much closer to their asking price if they asked for a more precise amount than a generic one (i.e. £396,500 rather than £400,000).

Eyes Right

The effectiveness of psychological pricing is also put down to something known as the 'left digit anchor effect'. This says that our eye is naturally drawn to the left-hand figure in the price so we make judgements based on that digit and largely ignore what comes after it.

Therefore we tend to make a decision on whether something is simply 'under £20 or over £20', rather than thinking about the whole price. Even if we are aware that retailers are deliberately creating these price brackets to tempt us, we still fail to make a logical analysis of the value of the item.

> **The pound sign (or whatever currency it is) put next to this digit helps add a certain credence to the number, emphasising its importance. Perhaps this is why we write £10, while saying it the other way round: 'ten pounds'.**

The Psychology of Pricing

One cunning consequence of retailers manipulating our brains using the anchor effect is that they can, theoretically, use this perception of 'value' to get away with charging us more than we would otherwise pay. If shoppers really don't go much beyond the first digit, why not charge £2.50 instead of the £2.00 you were going to charge?

As the missing penny, or indeed pound on more expensive items, falls further and further in value as the years pass, the psychological aspect of it only becomes more important to retailers. Despite credit and debit cards increasingly replacing cash, the symbolic, rather than practical, importance of that missing penny keeps increasing.

The technique may be as old as the hills, but don't expect it to disappear any time soon. As savvy shoppers we will continue to try and squeeze every penny we can out of retailers. Meanwhile retailers will continue to take advantage of our 'savviness' to squeeze those pennies out of us.

Incidentally, did you notice the cover price of this book is £9.99? Of course you did.

EAN13
CRACKING THE BARCODE

At 8.01 a.m. on 26 June 1974, a shopper named Clyde Dawson made history when he bought a ten-pack of Wrigley's Juicy Fruit chewing gum in Troy, Ohio. Mr Dawson's purchase was momentous because it was the first commercial scanning of a barcode or, to give it its full name, a UPC (universal product code) barcode. His small act was the culmination of three decades of work, which ultimately changed how people do business.

Reading a Barcode

The most recognisable barcode is the UPC, as first used by chewing-gum trailblazer Clyde Dawson. The black and white lines on this barcode are read by scanners to get information about the product. Every four bars or stripes on a barcode corresponds to one of the numbers below it, which saves time typing those numbers in.

The first six digits identify who the manufacturer is and the next five show what the item is. There is also another number on the right-hand end called the 'check digit', which is calculated from a formula using the other 11 digits to help the scanner work out if it has done its job properly.

There are different versions of the barcode, with the UK and many others, for example, using one called EAN13. This works in a similar way to the above with the code breaking down into: the nationality of the manufacturer (the first three digits), the company number (the next three to eight digits), the item reference (two to six digits), and then finally the check digit.

Scanning machines read the black and white lines based on their proportion to one another rather than their individual size – which is why you can blow up or shrink a barcode and it will still work.

Mad magazine once carried a giant barcode across its front cover with the message: 'Hope this issue jams every computer in the world.' It didn't.

A Striped Revolution

Checkout scanners do not read the price of items directly from the barcode but rather from a database called a 'price look-up file'. The barcodes represent identity numbers and these are sent from the scanner to the database to retrieve the price, which is displayed and printed on the receipt.

Once this is done the shop's computer systems will usually update the stock details to remind the company to order more of your favourite things. In this way, of course, the invention of the barcode revolutionised modern retailing. It wasn't just a neat way of recalling prices; it was the beginning of today's streamlined supply chains, a reliable method for shops to find out exactly what was selling in their store.

> **There is a conspiracy theory that guard bars – those bars that specify the beginning, the middle and the end of a number in many barcodes – represent '666, the number of the beast'. This is sadly untrue; these lines are not sixes and carry no information. Moreover you have to wonder why Satan would be interested in well-organised stock inventories.**

The Chequered History of Barcodes

The history of the barcode as we know it begins in the 1940s, when the owner of an American food chain approached the Drexel Institute of Technology in Philadelphia, asking for research into the development of a system that could read product information automatically at his supermarket's checkouts.

A graduate student named Bernard Silver overheard the request and in October 1949 he and a fellow student, Norman Woodland, patented a system using coded information – not in the form of the lined codes we know today, but in the form of concentric circles printed on paper.

But the idea was not an immediate hit in the US grocery sector and it sat around gathering dust for several years. Eventually a company called RCA, which had bought the students' patent, developed a similar bull's-eye symbol and scanner which it launched in a store in Cincinnati in 1972.

The Creation of Modern Shopping

Meanwhile in 1969 the American National Association of Food Chains approached the management consultants McKinsey & Co. for advice about creating a universal product marking system for their industry.

The big concern was that separate groups would develop different systems, which might be incompatible with each other. So McKinsey brought together the great and the good of the grocery industry to make up the 'Ad Hoc Committee of the Grocery Industry' – which, let's be honest, is a title that doesn't instill confidence that they were giving it their full attention. However, they came up with a smart solution, recommending a ten-digit code – five digits for the manufacturer and five for the product line – that would be printed directly onto the products by the manufacturer (this would be cheaper than printing them in each store).

All of these developments were noticed in Europe and in 1974 manufacturers and distributors in 12 European countries formed their own 'ad hoc' council (come on, guys, take it seriously!) to develop a similar system for Europe. By 1977 two different bodies had been established to administer their systems – one for Europe and a separate one for the UK, which was refusing as ever to play nicely with its European cousins.

> **Irving Nixon of IBM recalls seeing his first supermarket scanning lane in early 1976, which he says 'was built like a tank' and designed for a standing operator. It was also fiendishly expensive, costing £3,000 for the register and a further £3,000 for the scanner.**

Excess Baggage

The consultants at McKinsey persuaded retailers to adopt the use of barcodes by promising that it would create cost savings, but to start with this didn't really work out as they had envisaged. The consultants said electronic scanners would lead to a 40 per cent reduction in labour costs, because standing checkout operators could scan the goods and place them directly into a bag for the customer in one movement, thereby eliminating the need for a separate bagger. However, in practice many US supermarkets and their customers liked their baggers and stuck with them anyway.

In the UK, meanwhile, custom dictated (and still does) that supermarket customers should pack their own groceries – the service culture in Britain not being quite what it is in the US.

The Struggle for Barcode Supremacy

There were numerous other bumps on the road to barcode acceptance. For example, in the early days of scanning, direct sunlight through the large windows at the front of the shops dazzled the scanners and stopped the barcodes from reading. It was also discovered that barcodes printed red on a white background are invisible to scanners.

Then, in an act of rank snobbery, many wine companies refused to barcode their labels because they claimed the bottles were 'table decoration'.

But the purveyors of barcodes prevailed. The particular success of the European version led to it being renamed the International Article Numbering Association. However, the Europeans jealously guarded their part in this success and so to this day it is still known as EAN International, despite the fact it should really be IAN.

In October 1990, EAN International and its US counterpart signed an agreement to co-manage global standards together. Which was good news for us all.

> **All barcode numbers issued in Singapore start 888 and those from Korea start 880. Their barcode authorities secured these prefixes because 8 is a lucky number in Chinese culture.**

The Wonderful World of Barcodes

In recent years barcodes have become ubiquitous. *Poetry Review* once carried a poem consisting entirely of barcode fragments, which the poet claimed was 'machine readable'. There have been all sorts of barcode symbols developed since the 1970s to meet particular industrial applications, with the total now topping 300 variations. There are linear barcodes, and there are 2D versions that store lots more information. Hardcore enthusiasts, for example, will recognise Codabar, which is often used in libraries, medical facilities, photo-finishing and for airline tickets.

Everyone else won't.

> **The barcodes on newspapers differ from most others because they include the number of the day in the week as the second-from-last digit and the week number in a small additional code.**

562

THE PINT GLASS PUZZLE

The number 562 holds a special place in the hearts of many a drinker in the UK – even though most don't know it. That is because it is synonymous with a pint of beer.

This number, or one similar to it, is stamped on the side of pint glasses the length and breadth of the nation's pubs (and in the kitchen cupboards of many a light-fingered student). You'll usually spot it above or below a crown or another number signifying a year.

The mark proves that you are getting what you paid for – a pint. It is one of several aspects to the humble pint glass that make it considerably cleverer than you might first think.

What It's Not

Firstly, we need to put a couple of myths to rest. The most important one is that 562 is the number on all pint glasses.

It is not – there will be many of you for whom the number 2043 is a more common sight.

Secondly, the number 562 does not have anything to do with the 568 ml of liquid that makes up a pint, it's just coincidentally close. And where would we be if 'close' was good enough when it came to getting the right amount of beer?

Shifty Landlords and Short Measures

Since the dawn of time there have been landlords keen to serve you less than they should if it saves them a penny or two. In the late seventeenth century the English government decided to do something about it. To make sure drinkers were getting the correct volume of drink, they ordered stamps be put on drinking vessels.

Early on, this plan was no doubt as shaky as the drinkers' hands. But in 1879 the UK Board of Trade took another big step by issuing a circular to regional authorities telling them to adopt a uniform design of verification stamp 'with only variation of number or mark . . . as shall be sufficient to distinguish each Inspector's district'. The net was closing on the rip-off merchants.

In 1907 the regulations evolved to specify that a verification mark must 'be etched or sandblasted beneath or near the denomination outside the measure'. More recently the system became controlled by the nattily named Weights & Measures (Prescribed Stamp) Regulations of 1968 and various legal amendments that have followed it.

Calling the Beer Police

These days British law requires that draught beer (or cider) be served in measures of a third of a pint or a half of a pint or multiples of them. It's up to local authorities to make checks to see that manufacturers are creating glasses of the correct size.

The government issued each local authority with a series of stamp numbers and that authority in turn allocated these numbers to its men and women in the field. These brave souls are the Inspectors of Weights and Measures – the beer police, you might say.

These inspectors go to manufacturers and test their measures for accuracy and then, if all's well, pass them as fit for trade use. Of course they can't check every glass and only test a sample from each batch. Until 2006 their number had to be paired up with a crown to create a valid verification stamp, but under European regulations the regime now requires a 'CE' stamp and that the year of manufacture be stated instead.

Where Has Your Pint Glass Come From?

To regular drinkers it might appear that officers 562 and 2043 were particularly busy and conscientious, but that's not quite how the system works.

Stamps were allocated on a geographical basis, though they rarely remained in specific locations for a long time; in the past, numbers surplus to requirements in one area were reallocated to another part of the country.

A recent government list of all the available numbers reached 2,107 entries and stretched over 38 pages. But it also makes clear not all the numbers are still active. A large proportion of the digits you see appearing on your pint glasses are legacy numbers and once you've managed to drop and smash them all, then they'll be gone forever. Where a number has been cancelled, the name of the last weights and measures authority to which the number was allocated is shown in brackets.

With this in mind, let us take a quiet moment to think of our fallen pint glass protectors, including 240 (Grampian), 1005 (West Midlands) and 1408 (Croydon). You did many a drinker proud. We salute you.

So What's With 562 and 2043?

It's clear, then, that the ubiquity of the numbers 562 and 2043 on British pint glasses reflect the dominance of the glass manufacturers in these two areas. But how did this come to pass?

In short, technological advances have made production of glasses easy. Companies can only exist by making thousands of them at a time to keep the cost down. Many manufacturers have not survived this transition, or have given up on glassmaking altogether, leaving only a few specialists. One of the last centres of the British glassware industry was in Bury, north-west England – to which the number 562 was allocated. It had a virtual monopoly in the UK by 2006, when the European regulations forced its closure. That monopoly has ensured there are still 562 glasses all over the world and you'll see them around for a good while yet.

The other major development in the glassmaking world was that some trustworthy manufacturers won 'approved verifier' status, which means they can certify and stamp their own glasses. This brings us to stamp 2043, which belongs to the kings of the standard straight pint glass: Cristallerie D'Arques, part of JG Durand & Cie of Arques, situated just south of Calais in northern France.

They are extremely efficient and virtually no one else can compete with them. Hence the proliferation of 2043 stamped on your tipple. (So, yes, that is a French glass you are drinking Old Bulldog's English Pride out of – sorry.)

Will the Great British Pint Survive?

Of course, many ask whether there will be any role in the future for such numbers as the metric system continues its relentless march from Europe. The UK changed over to metric units in the 1990s with the final changeover taking place on 1 January 2000. Metric units of measurement must now be used for most transactions.

But here's the saving grace. 'The only imperial units permitted to be used for trade are the pint for draught beer, cider, and bottled milk and the troy ounce for precious metals,' the government says. And just to make sure: 'Imperial units may continue to be used alongside metric in dual labelling and consumers can continue to request imperial quantities.'

So the great British pint – and the assurance that you're getting 568 ml as required by law – remain safe for now. Cheers!

13

THE BAKER'S DOZEN

Those of us lucky enough to be given 13 loaves of bread when we've only requested (and paid for) 12 are the beneficiaries of an age-old tradition known as 'the baker's dozen'. The origins of this term are disputed, but the practice could go back as far as 1266, when the Assize of Bread and Ale law was enacted in England, dictating bread was sold by weight.

The thirteenth-century Assize of Bread and Ale law dictated:
  'By the consent of the whole realm of England, the measure of the king was made; that is to say: that an English penny, called a sterling round, and without any clipping, shall weigh thirty-two wheat corns in the midst of the ear, and twenty-pence do make an ounce, and twelve ounces one pound, and eight pounds do make a gallon of wine, and eight gallons of wine do make a London bushel, which is the eighth part of a quarter.'

Use Your Loaf

As the precise amount of food you got in the Middle Ages could be a matter of life and death, the penalties for ripping customers off could be brutal. Since bread was a staple for many societies, bakers were targeted with particular ferocity. Those who 'short-weighted' their customers were heavily fined, put in the stocks, or even flogged.

To protect themselves, they would sell 13 loaves for the price of 12, not least because accurate measurement was very difficult, meaning many an honest baker was no doubt punished for just making a mistake.

Historians note that originally the practice would have applied to any number of loaves sold, not just 12, because the law was based on total weight rather than total number.

> In fact, English bakers got off lightly; there are records of Egyptian bakers who didn't provide enough bread having their ears nailed to their bakery doors. In ancient Babylon, you could get your hand chopped off for the same crime.

The Assize law remained on the books for hundreds of years, only being repealed by the Statute Law Revision Act of 1863.

336

HOLES IN ONE –
THE GENIUS OF THE
GOLF BALL

As any golfer knows, each ball has a large number printed on it, usually just below the brand logo. This is purely for identification purposes – and for getting one over on arch-villains, like James Bond did in *Goldfinger*. But it may also contain other numbers that will probably be as much of a mystery to you as why you keep slicing that damned ball.

Show Us Your Dimples

Often there is a three-digit number printed on the ball, usually reading somewhere between 300 and 500. This is the number of dimples on the ball, which is important because those dimples make a big difference to how far you can hit it.

A ball without any dimples will go up to 150 yards or so if you hit it very, very well. The same shot with a dimpled ball could send it around 300 yards. The difference is down to aerodynamics: a smooth ball will shoot off the tee like a bullet, but will remain low and hit the ground more quickly.

Early golfers noticed that the original balls, which were smooth, would mysteriously start to go further once they became slightly damaged. They began to put their own nicks and scratches into the surface of their balls and the transition to dimpled balls began.

The Science Bit

What those golfers had discovered, even if they didn't understand the reasons behind it, was that a dimpled ball will spin (backwards if you haven't hooked or sliced your shot) and be lifted – or dragged – higher into the air, using much the same principles as are used to get an aeroplane off the ground.

The dimples also help minimise this extra drag by creating something called a 'turbulent boundary layer' of air round the ball. The upshot is a reduced area of low pressure behind the ball that would otherwise pull it back and impede its progress to the green. It is for these reasons you should try to keep your ball clean so the effect of the dimples is not distorted. (So feel free to hit it into a pond 'by mistake' now and again.)

> **There are no rules on how many dimples a golf ball should have, but somewhere between 300 and 450 is considered the optimum amount. The total number on any given ball will be down to the manufacturer, but 336 is particularly common.**

Ladies' Balls

If you've managed to keep hold of a ball long enough, it might also have another number, probably two digits, detailing the compression level of the ball. This shows how hard or soft the ball is, and is deemed important because when a ball is hit, it is partially flattened and then springs back into shape. If the compression level of a golf ball is not matched by the golfer's swing speed, it could result in either a lack of compression or over-compression – and ultimately lead to a loss of distance.

There used to be a serious machismo attached to this number because softer balls were considered 'ladies' balls', as you didn't have to hit them so hard for an effective shot. However, manufacturers tend not to make such a big deal of compression these days, not least because so many factors are known to affect a golfer's game other than the solidity of the ball.

To put it another way, you've got a lot more excuses these days for those forlorn trips to the bunker.

12
DAYS OF CHRISTMAS

There's a good chance that all the Twelfth Night of Christmas means to you is a last-minute dash to get the tree and decorations down before you are landed with bad luck.

But perhaps in a moment of peace, once the family have finally gone, you have wondered why there are 12 days of Christmas in the first place and why we celebrate Twelfth Night?

P-A-R-T-Y

The 12 Days of Christmas start on 25 December and end on the night of 5 January, traditionally with a great big party on the aptly named Twelfth Night. It marks the period from the Nativity (when Jesus was born) to the Epiphany, when the Wise Men arrived in Bethlehem to present gifts to the baby.

Incidentally there weren't three Wise Men in the original story, just three gifts; we have presumed that each was carried by a separate person and that no one forgot theirs or brought two and showed the other up.

> The word 'epiphany' comes from the Greek word for mani-
> festation, and was chosen because, to paraphrase religious
> historians, this was the night on which Christ – 'the King of
> the Jews' – was manifested (first appeared) to the Gentiles.

The Tree Graveyard

The 12 nights of celebrations are a very long-standing tradition. It was the Council of Tours in AD 567 that ruled that the period from the Nativity to the Epiphany would constitute one religious festival. (It also, unfortunately for Catholic priests, declared that any cleric found in bed with his wife would be excommunicated for a year.)

Sadly the Twelfth Night has lost much of its hedonistic sparkle and for many has just become the day their street turns into a graveyard for Christmas trees. But it used to be

a different matter altogether when, for hundreds of years, it was a night to let it all hang out.

Twelfth Cake and Lamb's Wool

One of the key elements to any good Twelfth Night party was the cake. Twelfth Night Cake – or just Twelfth Cake – was the forerunner of Christmas pudding and was baked with a bean and a pea inside it. When slices were handed out to revellers, whoever got the piece with the bean in it became King, while whoever got the pea was Queen. They then got to rule over proceedings for the rest of the night.

Of course any good party needs a little lubrication and traditional Twelfth Night celebrations were no exception. Booze was often distributed to guests from a ceremonial 'wassail', or toasting bowl, which was similar to a modern-day punch bowl. Traditionally it would contain a special drink, called Lamb's Wool, which was made from roasted apples, sugar and nutmeg in ale, or sometimes wine. It was called Lamb's Wool because when heated it would froth – or by some accounts 'explode'.

Out in the countryside (where they have always enjoyed a good bit of wassailing), the Twelfth Night would often be marked by the burning of twelve piles of straw. People would gather around the biggest one to toast the success of the harvest and have a drink. Or two. Or three.

It seems a shame that such a raucous festival has fallen from popularity. So, if you learn nothing else from this book other than to wassail as much as possible on 5 January, we will consider it a success.

419

MONEY FOR NOTHING

It is always nice to receive an email offering the promise of extraordinary wealth, gifts, prizes or employment. More often than not such messages come from a colourful band of folk, ranging from clairvoyants and lottery operators, to foreign government 'officials' and former African leaders.

'How nice of that deposed despot to think of sharing his blood-soaked millions with me – and first thing on a Monday morning too!' you might think. 'I see he wants a bit of money up front to release the cash, but that seems a pretty good deal in return for untold riches.'

Perhaps someone in Hong Kong has made you a fabulous job offer, which you can start as soon as you've transferred the funds to cover taxes, visas, 'anti-terrorism certificates' and a host of other items.

Of course most people realise these are scams, known in crime-fighting circles as Advance Fee Fraud, or 419 Fraud. Unfortunately there are still many, many people who are either foolish, desperate or just plain greedy, and fall victim to these villains.

Such ploys have been dubbed '419 scams' because many originate in Nigeria where section 419 of the Nigerian Penal Code prohibits such activity.

'One of the Oldest Swindles in the Book'

The practice, which was originally known as 'the Spanish Prisoner Letter', is as old as the hills. London's Metropolitan Police says it has been going on since at least the sixteenth century. A report in the *New York Times* in 1898, meanwhile, describes the scam as 'one of the oldest and most attractive and most successful swindles known to the police'. It goes on to tell of a 'well-known railroad man' in New Jersey who was contacted by one D. Santiago de Ochoa.

The letter said poor Señor de Ochoa was in jail in Cuba after deserting the Spanish army, but had fortuitously buried around $130,000 near the recipient's house when he was last in the US.

His daughter – so the story went – had a trunk that contained details of this buried treasure, but unfortunately she was under the protection of a 'hard-hearted boarding mistress'. To secure her release and gain access to the trunk could the railroad man please send money to . . .

Scamming the Scammers

There are some people who are taking the fight back to the

criminals, with several websites show-casing impressive efforts to engage and frustrate the fraudsters. In one instance a would-be victim persuades the Nigerian trying to defraud him that he will release money from a (fake) church fund if the man can show evidence of his religious zeal by getting a tattoo. Unbelievably the scammer does this to show his dedication to the 'Holy Church of the Tattooed Saint'.

A Deadly Joke

But while we may laugh at the absurdity of these offers and the fraudsters behind them, others won't get the joke. A lot of people have suffered after being drawn into such schemes. In June 1995, an American was murdered in Lagos, Nigeria, while pursuing a 419 scam. These are very nasty, very dangerous people we're talking about here, who scoop rewards estimated to amount to hundreds of millions of dollars each year.

It should be stressed that law enforcement agencies recommend that you should never have any contact with these people. If you do you are opening a very serious can of financial worms. And those worms can be very violent.

The best approach upon receiving these offers is summed up by the Metropolitan Police: 'If the promise seems too good to be true, it most probably is.'

+44

DIAL I FOR INTERNATIONAL

UK residents who happen to dial North America while also having a loose grasp of history might be struck by the thought that there is something amiss.

'How come the US got +1 as its international dialling code, while the UK only got measly old +44?' they might ask. 'Didn't one of our lot invent the phone in the first place?'

There is a certain sympathy to be had with that view. Indeed a Brit, or a Scot to be precise, Alexander Graham Bell, is widely credited with inventing the telephone in 1876.

But the fact is history, or nationalism for that matter, doesn't have much bearing on the international telephonic pecking order.

Setting Standards

It was the CCITT – or Comité Consultatif International Téléphonique et Télégraphique if you'd rather – that was behind the digits you dial for a foreign chat. This body is now known as the International Telecommunication Union, an agency of the United Nations that is charged with setting international standards for telecommunications.

The CCITT created the precursor of the modern system of international telephone country codes with its *Red Book* (nothing to do with Chairman Mao). This 1960 tome featured a list of country codes for Europe. It was probably never used by customers trying to make calls, as most were still connecting with each other through operators.

As a European directory (despite somewhat cryptically including several Middle Eastern countries and Russia), North America didn't even get a look-in. In fact, no one got the coveted +1 because 00–19 were reserved for 'special codes'.

> **The system was given an overhaul in 1964, with most countries already assigned numbers losing them. Only a few countries got to keep their original codes, including Greece (+30), Italy (+39), Sweden (+46) and the UK (+44).**

Into the Modern Era

The *Red Book* was superseded by the *Blue Book*, which divided the world into nine areas and, you guessed it, North America got area number one. The full list of World Zones goes like this:

1 – North America

2 – Africa

3 and 4 – Europe (it got two because there were so many
 countries)

5 – South America

6 – South-East Asia and Oceania

7 – Russia and Kazakhstan

8 – East Asia and others, including the Inmarsat satellite
 network

9 – Western Asia and the Middle East

Each country's code was thus determined by which zone it
was in. So, for example, all African codes start with a 2, such
as +27 for South Africa.

The African code also includes the newest member of the
international dialling community, South Sudan, which was
granted +211 in 2011 when it gained independence. The first,
historic, call on the number was made by the Undersecretary
in the Ministry of Telecommunications to his daughter in
Australia. Fortunately, it didn't go to answerphone.

Looking at the original *Blue Book* is also a reminder of
just how the world has changed in 50 years, listing, as it
does, North and South Rhodesia, West Germany and
Czechoslovakia, amongst other names that echo down from
history.

24

THE PUREST GOLD

The system of carats is used to measure the purity of gold. Carats are easily confused with . . . err . . . carats, which measure the weight of precious stones like diamonds. It's rather simpler in the United States where they're clever enough to measure gold in karats. But fear not, since gold has got its own weighting system – troy ounces – you shouldn't have too many worries the next time you get involved in a big bullion deal.

The Origins of the Carat

It is said 'carat' is derived originally from the Greek *kerátion* – 'fruit of the carob tree' – which evolved via the Arabic *qīrāt* and the Italian *carato* to give us today's word. Traders in the bazaars, souks and markets are said to have used carob seeds as weights on precision scales because of their reputation for having a uniform mass.

While carat may indeed have come from carob, a 2006 study led by Lindsay Turnbull from the University of Zurich found carob seeds have as much variation in their weights as other seeds. So if the story is true it may be that a lot of ancient people got ripped off when taking their valuables to be measured.

Pure Bling

In terms of gold, the higher the carat, the purer the gold. On the flip side of the coin, the lower the carat, the greater the proportion of other metals with which the gold has been alloyed.

Simply put, 24-carat gold is about as pure as gold gets. Completely pure gold has never been produced, although the World Gold Council says it is now possible to refine gold to 99.9999 per cent purity if you use the most up-to-date technology.

> **The measurement of gold purity in divisions of 24 has its origins in the fourth century. At this time the Roman Emperor Constantine minted a new gold coin that was the same weight as 24 smaller coins called siliqua – which happens to be the Latin word for carob.**

The Sliding Scale of Purity

With 24 carats being top-notch, gold alloy is then measured on a sliding scale as to what proportion of it is gold and what is not. So, for example:

- 18-carat (18K) gold contains 18 parts gold and 6 parts another metal or metals, making it 75 per cent gold.
- 14-carat (14K) gold contains 14 parts gold and 10 parts another metal or metals, making it 58.3 per cent gold.
- 10-carat (10K) gold contains 10 parts gold and 14 parts another metal or metals, making it 41.7 per cent gold. 10K gold is the minimum carat designation that can still be called gold in many places. Below that you have to call it 'gold-plated' or the like.

> **In Europe gold is also measured in terms of 'fineness', where 24 carat (99.9 per cent pure) is 999, 18 carat (75 per cent pure) is 750, and so on.**

Gold Rush

Of course, our modern-day obsession with 24-carat 'bling' is no recent phenomenon. In 95 BC, Chinese Emperor Hsiao Wu I minted a gold commemorative piece to celebrate the sighting of a unicorn. In a similar, slightly more realistic vein, Julius Caesar gave 200 gold coins to each of his soldiers as bounty from the spoils of war when he defeated the Gauls.

In more recent times, the California Gold Rush of 1848–55 provides a salutary lesson in the madness that greed can provoke. Some 300,000 people from all around the world descended on the area around San Francisco when a Mr James W. Marshall discovered gold on 24 January 1848, provoking one of the largest mass migrations in history. The prospectors were called '49ers' after the year many of them arrived. They faced innumerable dangers and hardships during their journey to California, not to mention illnesses and mining accidents once they arrived, and only a very few ever got rich. Marshall, who discovered the gold while building a sawmill, was forced from the land by prospectors and never became wealthy himself.

If you want to talk about a serious gold find, we recommend the largest-ever true gold nugget, which was found at Moliagul in Australia in 1869, weighing 2,316 troy ounces. It was called the 'Welcome Stranger'. The largest gold coin ever minted, meanwhile, was a 2007 Canadian $1,000,000 Maple Leaf, measuring a whopping 53 centimetres in diameter.

A Finite Resource

Given our endless lust for gold throughout history, why do we so often see it alloyed – or mixed – with other metals in jewellery? Well, the main reasons people want to debase gold are cost and malleability. Firstly, gold, as you are no doubt aware, is expensive. This is because there is not a lot of it out there. The World Gold Council says that only

166,600 tonnes of gold have been mined since the beginning of civilisation, and about 90 per cent of that has been found in the last 150 years or so.

> **If you were to pile all the gold that humans have pulled out of the ground into a box it would only cover twenty metres cubed. To put this in context, one ounce of pure gold could be hammered into a single sheet nine metres square.**

So mixing gold with other metals means you can maintain the yellow appearance, while making it more plentiful and cheaper at the same time.

Secondly, gold in its purest form is very soft and so is unsuitable for things like jewellery or coins. Alloying it with other metals means whatever you fashion it into is less likely to get broken or destroyed.

Toughening Up

The metals used to toughen gold up depend on the intended use. Jewellery, for example, gets a lot more handling than gold coins or bars and so is alloyed with harder metals such as silver, copper and sometimes nickel for the sake of durability.

The more expensive jewellery tends to be 18 carats (so 75 per cent gold and 25 per cent other metals) but comes with the caveat that you probably shouldn't wear it every day. Meanwhile bracelets and the like are more often 14 carats because of the constant wear and tear they suffer. Around 60 per cent of gold mined today ends up in jewellery.

> **Using different alloys can also change the colour of the gold; hence 'white gold', originally marketed as a cheap alternative to platinum, is made by fusing gold with alloys like nickel, zinc and palladium. Rose, red or pink gold (which is basically all the same thing) is created by upping the ratios of copper in the mix.**

Incidentally the gold bars that people always seem intent on stealing in films are known as Good Delivery bars. They weigh 400 troy ounces – about 12.5 kg – and the minimum purity required is 99.5 per cent gold. These are what professional bullion dealers in London, the centre of the world's 24-hour gold market, trade.

In fact, if you're foolish enough to be looking to steal some gold bars yourself, you could try American gold depository Fort Knox, which holds 4,600 tonnes of gold (the US Federal Reserve holds even more, at 6,200). Though given it is also the home of one of the world's most fortified military bases, you might want to think twice . . .

A4

VERY CLEVER PAPER

Somewhere near you, right now, there is A4 paper. It might be on a desk, in a pile in the corner, or perhaps floating past after an inopportune gust of wind. But it's there, you can be sure of it.

'So what?' you'll say. 'It's only a bit of paper, right?'

Wrong. It's actually an impressive piece of mathematics.

Almost every country in the world uses the 'A-sized' paper system (with the notable exceptions of the USA and Canada). Just imagine what would happen if everyone across the world used different paper sizes. The never-ending jams in photocopiers alone would be enough to drive many office workers to the brink.

A Certain Ratio

You might have noticed that A4, and indeed all the 'A' sizes from A0 to A10, have strange measurements. A4, for example, is 210 mm × 297 mm. This is because the dimensions of each size of paper are based on a formula that determines the ratio of the shorter side to the longer side.

This formula is simple: 1:√2 (one to the square root of two). Or if you'd rather, 1:1.41421356. What mathematicians discovered, as early as the mid-1700s, is that this ratio guarantees that if you cut a sheet in half through its longer side, the two sheets you are left with will always maintain that same ratio (i.e. they will be exactly half the size but stay in proportion at the same time). Try doing that with a square.

In modern terms this means that when using the photocopier you can change the size of a picture or diagram, from A3 to A4 for example, without distorting it or having to trim the edges to remove excess paper.

Early proponents of this sizing formula, such as eighteenth-century physics professor Georg Christophe Lichtenberg, saw the potential in this, while also enthusiastically commenting on how aesthetically pleasing the result was.

Just How Big is A0?

Paper sizes start with A0, which is one metre squared in total surface area. When the formula is applied we get dimensions of 841 mm × 1189 mm. Each time we go up a number, from A0 to A1 for example, the sheets of paper halve in size.

It then follows that A1 is eight times the size of A4, 16 times the size of A5, and 32 times the size of A6. The scale goes right down to A10, which no doubt has an inferiority complex, measuring just 26 mm × 37 mm.

> In case you're interested, there is also a 'B' range, which offers extra variations on its 'A' counterpart, as well as a 'C' range that is mainly used for envelopes.

Keeping Up Standards

The reason so many countries use these common paper sizes is down to a Berlin engineer called Dr Walter Portsmann. In the early 1920s he presented the idea of the 1:$\sqrt{2}$ ratio to the Standards Committee of German Industry, a forerunner of the International Organization for Standardization. The ISO is now the world's largest developer and publisher of International Standards.

Plenty of careful scrutiny and a whole load of paper cuts later, the formula became the German standard in 1922. Belgium got on the bandwagon in 1924 and it snowballed from there. Today the sizing is set in stone by 'ISO 216', which determines international paper sizing.

And here were you thinking that piece of A4 you just threw in the bin was nothing more than a sheet of paper . . .

3

POINTS FOR A VICTORY

FIFA's official line is that since it introduced its World Rankings of international football teams they have been 'a useful indicator for FIFA's member associations to find out where their respective teams stand in world football's pecking order and how they are progressing'. For countless football fans they have provided the chance to argue the merits of their respective teams until the sun comes up.

The Formula for Success

The ranking system was introduced in 1993 and given a major overhaul in 2006, after which a team's standing was judged on its performance over the last four years, compared to eight years previously.

Whether a team moves up or down the rankings is calculated based on several factors, which include:

- The **m**atch results (win, draw or defeat)
- The **i**mportance of the matches (competitive or friendly match)
- The strength of the opposing **t**eam (calculated using the team's position in the ranking)
- The regional strength, which is gauged by the results of teams from the respective **c**onfederation in the last three final competitions of the FIFA World Cup

These factors are multiplied together in the following formula to ascertain a team's total points (P). Each of the letters correspond to one of the characters in bold in the list above, giving us:

$$P = M \times I \times T \times C$$

'P' from all the team's games is then added together to give their place in the overall rankings.

Sounds Simple?

Before this formula can be applied you need to work out the respective value of each letter, which is more complicated than you might think.

This is how the system works in detail, courtesy of the good people at FIFA:

M: Points for match result

Teams gain three points for a victory, one point for a draw and no points for a defeat. In a penalty shoot-out, the winning team gains two points and the losing team gains one point.

I: Importance of match

This is allocated as follows:

- Friendly match (including small competitions): 1.0
- FIFA World Cup qualifier or confederation-level qualifier: 2.5
- Confederation-level final competition or FIFA Confederations Cup: 3.0
- FIFA World Cup final competition: 4.0

T: Strength of opposing team

The strength of the opponents is based on the formula: 200 minus the ranking position of the opponents. As an exception to this formula, the team at the top of the ranking is always assigned the value 200 and the teams ranked 150th and below are assigned a minimum value of 50. The ranking position is taken from the opponents' ranking in the most recently published FIFA/Coca-Cola World Ranking.

C: Strength of confederation

When calculating matches between teams from different confederations (the local governing bodies), the mean value of the confederations to which the two competing teams belong is used. The strength of a confederation is calculated on the basis of the number of victories by that confederation at the last three FIFA World Cup competitions. Their values are currently as follows:

- UEFA (Europe)/CONMEBOL (South America): 1.0
- CONCACAF (North and Central America, including the Caribbean): 0.88

- AFC (Asia)/CAF (Africa): 0.86
- OFC (Oceania): 0.85

> **In simple terms the biggest winners are the teams who win competitive matches, especially against high-ranking opponents. FIFA decided it couldn't just base the system on big matches like the World Cup because friendlies and small regional tournaments make up around 50 per cent of all the internationals that are played.**

Keep It Up!

The system also rewards teams that play regularly. Only results over the last twelve months count in full, those from the previous year count half, while games played up to three and four years earlier have even less significance.

Moreover, teams that play fewer than five matches in the preceding twelve months have their total number of points for that year divided by five. This means a team can't rest on its victory laurels.

A Rank Failure

The system was always going to have its critics since football fans like nothing more than a good debate about the beautiful game. However, some results that the rankings have thrown up in the past have surprised even the most objective commentators.

In 2008, two years after the system got its facelift, a bemused *Jerusalem Post* pointed out that 'despite failing to

THE SECRET LIVES OF NUMBERS

qualify for a major tournament in 38 years, the national team [Israel] is currently ranked number 16 in the world, ahead of such teams as Greece, Sweden and Denmark, and just two places below England'.

When questioned about this on a trip to Israel, FIFA President Sepp Blatter replied: 'I don't know why you're complaining. You're 16th in the world.'

Six months after the fall of Muammar Gaddafi the Libyan people got another lift when their team broke into the top 50 of the FIFA world rankings for the first time, in April 2012. The 'Mediterranean Knights', as they are known, jumped nine positions to 46.

10.10
WATCHES THAT SMILE

The next time you see an advert for a wristwatch have a look at what time it shows on the poster. Almost invariably the watch will be frozen at 10 minutes past 10, or alternatively 10 minutes to 2.

This has become the customary pose for most watch manufacturers for several reasons. Firstly, it frames their name nicely, as these are usually just above the centre of the watch. It also makes sure the hands don't obscure any extras like subsidiary dials or the date.

But the best reason of all is without doubt the aesthetic one: putting the hands at 10.10 makes it look like the watch is smiling, thus making it more attractive to the consumer. The last thing you want is a watch set at 20 minutes past 8, which looks like it has an attitude problem.

These posters are particularly prevalent at airports. You will now notice this 10.10 phenomenon a ridiculous number of times before you get out of the terminal. It will start to get annoying. Sorry about that.

7

DAYS A WEEK

The seven-day week is so ingrained in most cultures that anything else seems just plain weird. But trying to work out how we came to accept this as the norm is a different kettle of fish, because seven doesn't actually make a whole lot of sense in the scheme of things.

Since When?

There have been seven-day weeks reported from as far back as the Babylonian era, over 4,000 years ago. There are other stories that give bragging rights to the Persians, the Greeks and/or the Jews. The passage of time has made it pretty tough to get a definitive answer on who was using it first.

There are claims, which are very difficult to substantiate, that Sargon I, King of Akkad, was first to impose a seven-day week around 2350 BC. (The only thing we can be sure of is that it's just a matter of time before a film producer uses his name as an intergalactic despot in some big-budget sci-fi movie: 'I am Sargon, fear my wrath!' etc.)

What can be said with much more historic certainty is that the idea went mainstream in the fourth century when the Roman emperor Constantine adopted it. He replaced the traditional eight-day week with a seven-day version because, as the first Christian emperor, he was keen to promote the religion's traditional cycle: six days, then a day of rest.

Why Was Seven Chosen in the First Place?

It seems ancient man created the concept of weeks to bridge the gap between a day and a month, which was not an unreasonable thing to do. However, most of our concepts of time are based around the movements of the sun and the moon – and seven doesn't really fit with that. For example, the lunar cycle, from start to finish, is about 29.5 days, give or take a decimal place.

'Four times seven gives us 28 – that's pretty close,' you might say. This has often been proposed as the background to the seven-day week: it was an act of expediency by astronomers. Also, if you were to do five six-day weeks, you overrun the lunar cycle, thus making seven slightly more attractive.

But given the astonishing accuracy of even ancient astronomers, one has to wonder whether we are doing them a disservice by claiming they would put up with such a slapdash approach to chronology.

The Planets Have It

The other theory that holds a certain amount of sway is actually pretty straightforward. It says that the only large astronomical bodies visible to the naked eye of ancient man were what we now call the Sun, Moon, Mars, Mercury, Jupiter, Venus and Saturn (words we inherited from the Romans).

So, how many's that? Oh right, yes, of course: seven. No one was quite sure what these planets were up to, twinkling away and moving slowly across the sky. But early man was pretty sure it was something divine. In fact, our word for 'planet' comes from the Greek *planetes* – or πλανήτης if you want to be all exotic about it – which meant 'wanderer': a celestial being, or god.

The Days Belong to the Gods

The story goes that each 24-hour day (a concept itself invented by the ancient Egyptians) was given a ruler, in the form of one of these seven heavenly bodies. After the seventh stellar deity had had their day in charge, the first would take over again.

In the English language, Sunday remains a pretty straightforward reminder of this, as do Monday (Moon day) and Saturday (Saturn day). The other days of the week aren't quite so clear because you have to run them through a historical and cultural decoder first. Tuesday to Friday were given to us by the Saxons, who originally came from northern Germany and south Denmark. They brought the Norse gods with them, including Tiw, Woden, Thor and Frigg, after whom they renamed the planets, and hence the days. The Norse days of the week stuck with us, but the Saxon planet names eventually reverted back to their Roman predecessors.

Other countries weren't so bothered by marauding Saxons and their Norse naming habits. Thus France has *mercredi*, Spain *miércoles* and the Italians *mercoledì*, all of which are 'the day of Mercury'.

Incidentally, the Norse god Frigg, who gave us Friday, was the goddess of love and beauty – which makes one wonder if ancient workers looked forward to the end of the week as much as we do.

Is Seven a Sure Thing?

While there are examples of various cultures still using different lengths of week around the world to this day, seven has become so ingrained in much of the human psyche that it seems certain to remain the dominant global measurement for the foreseeable future.

But centuries of seven holding sway over the week haven't stopped successive regimes trying to alter it, right into the twentieth century. Two notable examples are the French and the Russians.

The Ten-Day Week

The French came up with the Republican Calendar in 1793 as the revolutionaries tried to purge religious influence and charge headlong into decimalisation. This overhaul left poor

French workers toiling three ten-day weeks a month, getting the final day of each week as a day off.

Unsurprisingly, this didn't go down well, since they had previously only had to work six days in order to get a holiday. The system lasted until the end of 1805, when Napoleon abolished it.

The Five-Day Week

The Russians, meanwhile, tried all sorts of techniques to improve the calendar, largely to get more productivity out of the workforce. Between the autumn of 1929 and the summer of 1931, the government experimented with a five-day week, which saw the calendar year divided into 72 portions. Random holidays or (more likely) work days were then added to balance the books.

Every worker had one day off a week, but there was no fixed day of rest. This meant your break could be different to the rest of your family and friends, which must have led to some pretty boring – or restful – days off, depending on the state of your relationships.

This burden on the comrade brothers and sisters was remedied when the Soviet calendar was changed in 1931 to a six-day week with a fixed day of rest at the end of it. This system lasted until June 1940 when normal service was resumed with a seven-day stint, by which time many Russians were at war and probably getting no days off at all.

E102
THE BAD BOY OF E-NUMBERS

Food additives, also known as 'E-numbers', are in much of the food and drink that you find almost anywhere you choose to shop, from supermarkets to delicatessens. It is often said that they are to be avoided, particularly by children, unless you want a bout of crazed hyperactivity, a nasty headache, or your skin to turn a funny colour.

They are included for a whole host of reasons, such as:

- To preserve food (giving them greater shelf-life)
- To add colour or thicken a product (i.e. to improve the look and feel of it)
- To add sweetness
- To enhance flavour

As a general rule, colouring additives are given numbers 100 to 181, the antioxidants are 300 to 340, flavour enhancers cover 600 to 650, and glazing agents sit in the 900 to 910 range. Beyond that, up to E999, the various items fall into the 'miscellanous' bracket.

The Yellow Peril

The 'E' stands for Europe after it became a legal requirement in the 1980s to detail any additives in food sold in the EU. This means they have been tested and given the thumbs-up in terms of health and safety. How safe they actually are is a matter for some pretty heated debate, particularly if your child starts careering round like a rogue missile after downing a soft drink.

The E-number that is the poster child for unhealthy additives is E102, or 'tartrazine', which has been linked to asthma, hyperactivity and rashes. This yellow dye is all over the place, unfortunately. It is in everything from drinks, soups and jams, to many convenience foods. When it is combined with blue dyes you'll even find it in cans of peas. Ironically, if you're intolerant to it then you might find yourself in a pickle when taking a cure, because tartrazine is often used to colour medicinal capsules. (That's clearly a joke – don't take medical advice from this book, whatever you do.)

E-facts:
> **E951 is aspartame, an artificial sweetener that can be up to 200 times sweeter than sugar.**
> **E901 is made of beeswax and is used for glazing and polishing food.**
> **E903 is carnauba wax, which is often used for shining up confectionery. It is also used for polishing cars.**
> **E948 if you want a breath of fresh air then this is the one for you – it's oxygen.**

When E-Numbers Are Good for You

It isn't true to say E-numbers are all bad. While E102 and its kin are artificial additives, others are just numerical representations of perfectly ordinary natural products. These include E300 (vitamin C) and E162 (beetroot juice).

There are also 'nature identical' additives that have exactly the same chemical composition as natural substances, even though they are man-made. (These compare with their artificial counterparts that don't share their chemistry with any naturally occurring matter.)

> **Anyone looking for a bit of bling with their food should look out for E174 and E175, which are colourings derived from real silver and gold (although this does technically mean you are eating metal).**

9 x 9

SQUARES IN A
SUDOKU GRID

The first thing you need to know about the game of Sudoku is it didn't originate in Japan. Now, hold on to that thought while we explain how the game works, before delving into its globetrotting history.

The Number-One Number Puzzle

For those who haven't played Sudoku (you're missing out, by the way), the rules are simple, even if applying them is not. The objective is to fill a nine-by-nine grid so that each column, each row, and each of the nine three-by-three boxes (also called blocks or regions) contains the digits from one to nine, but only once each. The puzzle setter provides a partially completed grid, leaving a certain amount of numbers for you to work with.

For a Sudoku to be the real deal then there can only be

one solution, unlike in 1995 when UK Sky TV carved a giant Sudoku into a hillside as a publicity stunt. It turned out the 84m^2 puzzle had 1,905 solutions, making it pretty tough to work out who should win the £5,000 prize money.

There are 6,670,903,752,021,072,936,960 possible unique combinations for completing a standard nine-by-nine square Sudoku grid.

The Globetrotting History of Sudoku

You could, if you so wished, trace the game back thousands of years through Arabia to China in the form of 'Magic Squares'. These were puzzles wherein a square grid was filled with numbers in such a way that each row, each column, and the two diagonals added up to the same number.

One ancient example is based upon a Chinese legend involving a river god, a boy and a turtle. (Intrigued? Look it up, there isn't space here to explain.) It has three rows and three columns, and if you add up the numbers in any row, column or diagonal, you always get 15:

4	9	2
3	5	7
8	1	6

Jumping ahead several millennia to the eighteenth century we meet the extraordinary Swiss mathematician Leonhard Euler, who is credited with being the Godfather of Sudoku. In 1783 he came up with a grid in which no number appeared twice in the same row or column. It was called a Latin Square because of Euler's use of Roman symbols.

Again, using this example, you'll see the difference between it and a Sudoku: it isn't subdivided into the three-by-three 'blocks' containing unique numbers:

1	2	3	4
2	1	4	3
3	4	1	2
4	3	2	1

There were several versions of this presented for puzzlers as time marched on, including in late-nineteenth-century French newspapers, but it wasn't until the late 1970s that Sudoku emerged as we know it today.

Great Idea – We Might Need to Work on the Name . . .

In May 1979, New York-based Dell Magazines published the first recognisable Sudoku in their *Dell Pencil Puzzles and Word Games*. Known as 'Number Place' in those days, the game was most likely designed by Howard Garns. This retired American architect from Indiana made his mark on history by taking the Latin Square concept and adding the nine three-by-three 'blocks' to enhance the puzzle.

The concept was embraced by Japanese puzzle company Nikoli, who began publishing them in 1984 under the tongue-tying moniker '*Sūji wa dokushin ni kagiru*' – or 数字は独身に限るto be precise – meaning 'the numbers must occur only once'. It didn't take long for frustration, bewilderment, or perhaps just laziness, to ensure it was shortened to Sudoku – single (*su*) number (*doku*).

Funnily enough, these days the puzzles are often called by the English name, Number Place, in Japan; whereas in English publications, they often go by the Japanese name, Sudoku.

In 1986 game company Nikoli added two new rules that said the most 'given' numbers in a puzzle could be 32 and that the puzzles must be symmetrical – i.e. the boxes with given numbers in them are the same if the puzzle is rotated 90 degrees.

World Domination

The game took Japan by storm but the numerical riddles largely remained a mystery to the rest of the world for the next two decades. Then one day Wayne Gould, a retired judge from New Zealand, was browsing a bookshop in Tokyo and spotted one of the puzzles.

He found himself so compelled to fill it in that he spent the next six years developing a computer programme that created ready-made Sudokus. He took the concept to the

UK *Times* newspaper, where editors showed considerable foresight by taking the idea on, publishing the first one in November 2004.

A few months later the puzzles were everywhere, including New York where Sudoku's modern evolution began. In 2006 they had gained so much popularity that the first Sudoku World Championships were held.

With thousands of years of history and some amazing twists and turns behind Sudokus, the question remains: are you more intrigued by the story of the boy and his turtle?

P165
THE CAR TYRE CODE

You may have noticed that there is a code printed on your car's tyres, which will read a bit like this: P165/65R14 79T. Now, solving this puzzle won't lead you to hidden treasure – but it may well save you a bit of cash, as well as enhancing your driving experience.

Cracking the Tyre Code

The code mentioned above is a mine of information, including your tyre's width, profile, rim size, load rating and speed rating. The example mentioned above, P165/65R14 79T, tells us the following:

- P: This letter, which may or may not be there dependent on the make of tyre, shows its intended use. In this case, 'P' is for passenger vehicle.
- 165: This is the width of the tyre measured in millimetres.

The wider it is, the better the grip, but the worse the fuel economy and the noise.

- 65: This is the 'aspect ratio' of the sidewall height to the total width of the tyre, as a percentage. The shorter the aspect, the better for handling – that's why smaller-looking tyres are often seen on sports cars. In this case, the height of the sidewall is 65 per cent of the size of the 165 mm width.
- R: This means 'radial construction', detailing how the tyre was built.
- 14: This is the diameter of the tyre's inner rim in inches.
- 79: This is the load rating of the tyre, a code that represents the maximum weight a tyre can carry. The '79' is the code for a total load of 437 kg.
- T: This details the speed rating, i.e. the maximum speed the tyre can handle at full load without becoming damaged. At one end of the scale is Q (99 mph), while Y tops out at 186 mph. In this case, T, the tyre can travel safely at up to 118 mph or 190 kph, but this is only for a limited period, so best not to push it.

Peak Performance

It's well known that driving with car tyres properly inflated can make a big difference to fuel consumption and performance, but people are less aware that having the right tyre in the first place also helps a lot. Having the correct set of wheels can improve stability, response times and fuel consumption. If you think about it, at any one time the only thing keeping you attached to the road is four pieces of rubber. So the importance of wheels to your driving experience is never to be underestimated.

666

THE NUMBER OF
THE BEAST

The number 666 strikes fear into the hearts of many: it is the 'number of the Beast' and its appearance is often said to herald the end of the world.

There are some very strong opinions out there about what 666 refers to and who or what the 'Beast' is. There is also furious argument over whether the Beast is long gone, with us at the moment, or yet to arrive.

Former American President Ronald Reagan may have held his nerve, one finger poised over the big red button, during the Cold War, but nuclear conflagration was clearly not as scary as his retirement address of 666 St Cloud Street. The Reagans had this changed to 668 (which, you have to suppose, made them 'neighbours of the Beast' instead).

The central point of reference for 666 is the Book of Revelation, which is found right at the end of the Bible. The bit you're looking for is Chapter 13 Verse 17, which goes like this: 'This calls for wisdom. If anyone has insight, let him calculate the number of the beast, for it is man's number. His number is six hundred three score and six.'

That number is of course an old fashioned way of saying 666.

The Rising of the Beast

So why the horror? Well, the passage that precedes it goes some way to explaining why people get upset. It describes the appearance of the Beast and here is an abbreviated version:

> I stood upon the sand of the sea, and saw a beast rise up out of the sea, having seven heads and ten horns, and upon his horns ten crowns, and upon his heads the name of blasphemy . . . And I saw one of his heads as it were wounded to death; and his deadly wound was healed . . .
>
> And it was given unto him to make war with the saints, and to overcome them: and power was given him over all kindreds, and tongues, and nations . . .
>
> And I beheld another beast coming up out of the earth; and he had two horns like a lamb, and he spake as a dragon . . . And he exerciseth all the power of the first beast before him, and causeth the earth and them which dwell therein to worship the first beast, whose deadly wound was healed . . .

And he causeth all, both small and great, rich and poor, free and bond, to receive a mark in their right hand, or in their foreheads: And that no man might buy or sell, save he that had the mark, or the name of the beast, or the number of his name.

An Unfortunate Birthmark

Towards the end you'll spot the reference to the 'mark of the Beast', which, if you've seen the film *The Omen*, you'll know is actually a 666 birthmark on a scary-looking child. There are some quite extraordinary (or worrying, dependent on your viewpoint) interpretations out there about how the mark is, or will be, manifested. These range from the number itself, to tattoos, barcodes and microchips.

There is one theory that says the Internet is the mark of the Beast because in Hebrew 'w' is also the same character as 6 – so the 'www' prefix on website addresses equates to 666. But no matter how much weird stuff there is on the Net this doesn't stand up to scrutiny because it's three sixes placed next to each other – rather than the 'six hundred three score and six' quoted in Revelation.

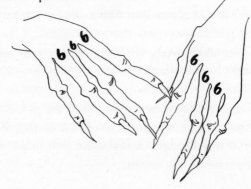

A Warning to the Early Christians

Most people are familiar with the idea that the number 666 is associated with the coming of the Antichrist and visions of the Apocalypse. But putting aside hellfire and brimstone for just a moment, there is a fascinating explanation of this prophecy rooted in the day-to-day life of the time the passage was written.

The Book of Revelation is thought to have been created some time towards the end of the first century AD. This was an especially rough time for early Christians, with one man in particular standing out as their chief tormentor.

Emperor Nero ruled the Roman Empire from AD 37 to 68. After Rome burned in AD 64 the finger of blame was pointed at Nero (who was allegedly fiddling at the time). He in turn tried to blame the Christians, who were a minority religious sect in those days. The persecution that followed was horrendous.

It was in this context that the Book of Revelation was written – as a letter of encouragement to Christians suffering under Roman rule. Many historians say it is worth interpreting all the references to 'the Beast' with this in mind. They say that the Beast represents the Roman Empire with its horns and heads as symbols of the rulers of Rome.

Now here's the key bit from the description of the Beast: 'I saw one of his heads as it were wounded to death; and his deadly wound was healed.'

This is supposed to be a reference to Nero, who committed suicide in AD 68. The big fear after this would have been that the hammer of the Christians could come back. Even today the deaths of famous people throw up all sorts of conspiracy theories (Elvis seen working in local supermarket, etc.). It's not hard to imagine John, the author, advising early believers to keep an eye out for Nero, in case he had simply been in hiding, biding his time before a bit more Christian bashing.

The Code of the Beast

This takes us neatly to one explanation as to why John chose 666 as his beastly number: it is a code. In Greek (which is very possibly what the Book of Revelation was originally written in), Nero's name is 'Kaiser Neron'. This would be represented by the Hebrew letters qsr nrwn, which also functioned as numbers:

1) $q = 60$
2) $s = 100$
3) $r = 200$
4) $n = 50$
5) $w = 6$

So we see qsr nrwn = $60 + 100 + 200 + 50 + 200 + 6 + 50 = 666$

There is a problem with this in that many sources say the original number of the Beast was 616 and it was later changed to 666 to make it more memorable. The proponents of the code theory reply that this is because some early scribes dropped the final 'n' on Neron, losing 50 points in doing so.

Other academics applying a similar system say it actually fits Emperor Caligula, who offended the Jews in particular by planning to build a statue of himself inside the Temple at Jerusalem.

However, if you are more of a fan of the notion that a sighting of the number 666 will bring with it Satan, hell-fire and general damnation, then when the Four Horsemen of the Apocalypse ride into town do feel free to say, 'I told you so.'

10

THE FAMOUS BLACK DOOR

Number 10 Downing Street is one of the most famous addresses in the world. It has been home to British Prime Ministers since 1735 and sits amongst some of the most expensive real estate on the planet. But it hasn't always been that way.

'The Terrible Place'

The earliest records name the site as Thorney Island – literally an island covered in thorns – and describe it as a marshy and unpleasant area. It only thrived because it was there that a ford across the River Thames joined a Roman road, making it a transportation hub.

But the land was boggy, creating problems for building that have hampered the site right up to the modern day. The marsh also made it an ideal place to contract bubonic plague.

The official 10 Downing Street website records that, unlike now, the area's inhabitants were originally very poor. It goes on to detail a document of the Mercian King Offa in AD 785, which names it 'the terrible place which is called Thorney Island'.

But from the start of the eleventh century a series of kings (namely Canute, Edward the Confessor and William I) put their faith in the place, building castles there. This made the area more desirable and no doubt easier for medieval estate agents to sell.

The Axe Brewery was the earliest building known to have stood on the site of Downing Street, but by the early 1500s it was no longer in operation. The first domestic house we know about was leased to Sir Thomas Knyvet by Queen Elizabeth I. He was a Member of Parliament for Thetford and Justice of the Peace for Westminster who won his place in history by arresting Guy Fawkes, the rebel Catholic about to blow up the Houses of Parliament in the 1605 Gunpowder Plot.

Shaky Foundations

The man who gave his name to this famous address was far from upstanding. Sir George Downing moved from spy to traitor to property developer with consummate ease. Famous diarist Samuel Pepys called him 'a perfidious rogue'. Over 200 years later Winston Churchill described him as a 'profiteering contractor'.

Downing realised that the area was a potential property goldmine because it was so close to the heart of the

Establishment. Both Parliament – the centre of government – and the headquarters of the Anglican Church at Westminster Abbey were just next door.

In 1654 Downing bought the land but it took almost 30 years to build on it because descendants of Knyvet (he of Guy Fawkes fame) had a lease on it. But in 1682 that lease ran out and Downing began creating his now famous street by pulling down the existing buildings and (almost literally) throwing up a small cul-de-sac of 15 to 20 houses.

The new buildings were cheaply built on bad foundations – Downing even had mortar lines drawn on the front of them to make it look as if they had attractive brick facades, when they had anything but.

Number 10 did not actually gain this title until 1779. Before that it had been Number 5. Adding to the confusion for the poor postman, correspondents tended to ignore the numbers in favour of naming the house after its occupants.

Two Houses Become One

The current Number 10 is made up of more than Downing's house. It is two homes joined together: the cheap terraced house and a former aristocrat's grand abode behind it.

The latter looked out over the impressive Horse Guards Parade and Hyde Park. But in a move that can only be described as 'British', the leaders of what became the most powerful nation on Earth decided that the crumbling terraced house should be used to welcome honoured guests.

Living Above the Shop

Number 10 first became the residence of a prime minister in the 1730s when King George II gave both houses to Sir Robert Walpole, who held the title First Lord of the Treasury. Being a generous soul, Walpole refused the gift and asked the King instead to make it available to him, as well as future First Lords of the Treasury, to use as an office or home. This explains why that title is still engraved on the brass letterbox on the front door.

For years prime ministers rued Walpole's gift because of the structural defects in the building and many chose not to live there if they could avoid it. In 1782 the Board of Works recommended that 'no time be lost in taking down said building'. These problems have continued throughout Number 10's life and have led to repeated building work on the site, including a major rebuild in 1960.

But the nineteenth century brought new problems as the area became a haven for crime, prostitution and other underhand professions ('such as politics', some might say).

Benjamin Disraeli noted the house was 'dingy and decaying' when he moved there in 1868. It was not until the turn of the twentieth century, when Prime Minister Arthur Balfour lived there, that the house became the consistent residence of the PM. These days the prime minister's family resides on the second floor; 'living above the shop', Margaret Thatcher called it.

The Most Famous Door in the World?

While the history of Number 10 is long and distinguished, the most famous part of the building is not the grand state-rooms inside, but its exterior.

The front door is more than a door – it is an icon. It is a place where countless world leaders, suppliants, campaigners, celebrities and many others have stood before entering the corridors of power.

> It is also more than 'just a door' in other senses. If you look carefully you'll see the 0 of the number 10 leans to one side – this is a wry nod to the 0 on the original door that was mounted badly.

You might also have noted the door is peculiarly shiny. This is not simply because it is regularly cleaned but rather because the door is made of bomb-proof metal. The original oak door was replaced in 1991 after a mortar attack on Downing Street by the IRA. Whenever the door needs to be refurbished it is removed and replaced with a replica. It is so heavy several people have to lift it.

Look around the door and you will note the famous black brickwork, which isn't really black at all. In the 1950s the street was spruced up and the cleaning work revealed the black bricks were in fact yellow, but had been sullied by 200 years of pollution. The dark facades were deemed so recognisable that the decision was taken to paint them black again.

Ding-Dong?

If you are lucky enough to gain admittance to Downing Street, having run the gauntlet of police, security guards and baggage checks, you are met with an eerie calm in this small street that is often empty of people.

A word of caution. When you approach the famous address, don't bother ringing the doorbell; it doesn't work. Don't worry – you can be pretty sure someone will be watching you on the monitor inside.

Of course 10 Downing Street isn't the only famous London address. Number 221b Baker Street is another iconic landmark, known to millions as the home of 'consulting detective' Sherlock Holmes and his partner, Dr John Watson. They lived there between 1881 and 1904, according to the stories by Sir Arthur Conan Doyle. It is now the site of the Sherlock Holmes Museum, which maintains the first floor study in Victorian period style.

However, there has been quite a row over who has the right to receive the great detective's mail. The museum was only given the address in 1990. Its original address was 239, whereas 221 was home to the Abbey National building society. The latter used to employ someone to answer mail addressed to Sherlock Holmes.

Famous Addresses

- 4 Privet Drive, Little Whinging, is where Harry Potter grew up in J K Rowling's books. According to the Pottermore website, Rowling chose '4' for the Dursleys' address because she found it 'a hard and unforgiving number'.
- 32 Windsor Gardens, London, is where Paddington Bear lived having left Deepest, Darkest Peru.
- 1600 Pennsylvania Avenue, Washington, is the home of the President of the United States.
- Number 1, London, otherwise known as Apsley House, is the former London townhouse of the Dukes of Wellington.
- 6 Rue Chantereine is the original Parisian address of one Napoleon Bonaparte, and the house from which he launched his coup d'état in 1799. The street was renamed Rue de la Victoire – and the house renumbered 60 – as a tribute to Napoleon's military triumphs.
- 10236 Charing Cross Road, Beverly Hills, is home to the Playboy Mansion.

22

RUNS OFF ONE BALL: THE DUCKWORTH–LEWIS METHOD

This mathematical formula was created to decide who should be the winner when one-day cricket matches are interrupted by rain. Unsurprisingly it was invented in England where rain features almost as often as the cricket itself.

If the heavens do open, the Duckworth–Lewis Method can be applied to the state of play to establish what the teams need to achieve in order to win. It is named after Frank Duckworth and Tony Lewis, the statisticians who designed it.

The concept of the Method is quite straightforward. The mathematical chicanery that goes on behind the scenes is anything but. It has often been said that to understand the workings of the Method you don't need a degree in astrophysics – but it helps.

Rain Interrupting Play

The origins of the formula lie in the fiasco of the 1992 Cricket World Cup semi-final, when rain interrupted play with South Africa needing 22 runs off 13 balls to beat England. This was an eminently achievable target. When the rain stopped, the system in place at the time decreed that South Africa still needed 22 – but off a single ball. Seeing as six runs is the most that can be scored in one ball in cricket, this was a disaster.

Frank Duckworth recalls a commentator on the radio lamenting that 'there must be a better way than this' and so he and Tony Lewis set about resolving the problem. Duckworth says an idea on an airplane to Hawaii and a modification from Lewis led the pair to realise they'd finally cracked it.

Making Use of Your Resources

What they came up with was a system based on the notion that teams have two resources with which to make as many runs as they can. These two resources are the number of overs a team has still to receive and the number of wickets that team has left. The combination of these two factors will, at any stage in their innings, determine the ability of the players to score more runs.

It then follows that if rain shortens a game the resources available to one team are reduced. So for example, if the team that opens the batting has its innings interrupted, then its players have had some of their ability to score taken away. So the second team to bat would usually be set a larger run target than the first team scored to make up for this.

When the opposite happens and the second team to bat is interrupted, their run target would be reduced. In either case there is a need for revised targets to be set so the game can be played out in a way that is theoretically as close as possible to what would have happened if it hadn't rained.

Are you with me so far? OK, let's get into the meat of it.

Rainy-Day Reading

There is a published table of 'resource percentages', which is basically a table with the number of wickets lost detailed along the top and remaining balls and overs down the left-hand side.

The full table covers each individual ball in a game of up to 50 overs per side, while there is also a simpler over-by-over version of the table. Both allow the players to effectively calculate the percentage of their batting resource that has been lost on account of play being suspended.

So before the game where there are 50 overs still to be bowled and no wickets have been lost, the resource percentage available will be 100 per cent. This percentage is reduced the more overs are gone and the more wickets have fallen.

Calculators at the Ready

Messrs Duckworth and Lewis say that their method can be applied using this table and a pocket calculator (although there is a more complicated computer pro-gramme, the Professional Edition, which is usually used in the top flight).

THE SECRET LIVES OF NUMBERS

It might seem complex but the Method has been in service for 15 years now. It was first used in the 1996–7 cricket season and officially adopted by the International Cricket Council in 2001. It has been generally welcomed by the cricket community and everything from a horse to a music group has been named after the mathematical duo.

Ducking Criticism

But it has also come in for its fair share of criticism. It has been described as everything from 'bizarre' to 'duck soup'. There are technical arguments over it, such as the weighting of wickets to overs, and there are more romantic complaints around whether mathematics should be applied to sport in this way. If, as a cricket fan, you don't like it then the challenge is to come up with something better. But you'll probably need to go and get that astrophysics degree first.

Incidentally, if the Method had been applied to the 1992 World Cup game that inspired it, South Africa would have required an eminently achievable four runs off that one ball rather than 22. What a finish that would have been.

555

THE TELEPHONE CODE
THAT DIDN'T EXIST

There's a good chance that even if you're not American then the US telephone code of 555 will ring a bell somewhere in the back of your mind. This is because 555 has been used as a fictional dialling code in numerous films and TV shows since the 1960s.

The numbers shown in these programmes don't work if you call them, meaning crazed fans won't bother innocent bystanders while trying to track down their fictional heroes.

Not all 555 numbers are fake: 555-1212 is one of the standard numbers for directory enquiries in the USA and Canada, for example. In fact these days only numbers from 555-0100 to 555-0199 are now reserved for fictitious numbers. The 555 code followed earlier US TV programmes of the 1950s and 1960s, which often used 'KLondike 5' in a time when telephone exchanges used letters and numbers.

> In one episode of *The Simpsons*, Police Chief Wiggum dismisses a number his officers have traced because it contains 555. 'Aw, geez, that's gotta be phony,' he tells them.

Dial D for Drama

It's not only the US that uses fake codes in this way. If you want to stage your own blockbuster it's easy to go online and rent telephone numbers that have been set aside by local regulators.

British telecomms body OFCOM has 1,000 'UK wide numbers' officially reserved for drama that run between 03069 990000 and 03069 990999. You can even select your city of choice. If you see someone calling a Liverpool number on your TV, for example, they will dial one of the 1,000 drama numbers officially reserved between 0151 4960000 and 0151 4960999. If your production demands a bit more sun and would benefit from a shot of the Sydney Opera House then the Australian Communications and Media Authority has 9,000 drama numbers officially reserved (between 02 70101000 and 02 70109999).

Phony Phone Lines

Some productions have used real numbers by mistake in the past, while many more have cottoned on to the idea of using them as a marketing tool. In one episode of the medical comedy *Scrubs*, viewers strange (or bored) enough to try calling one of the doctors' phone numbers apparently heard a recorded message from the cast.

One company, Fictional Telecom, offers unique drama numbers that are fully operational, which the firm suggests filmmakers might use to give fans:

- A voicemail greeting of your fictional character
- A recorded announcement promoting your book or movie
- Interactive phone menus and announcements
- Continuous ringing or a busy/not in service message

Of course a further option is to rent a line if you are trying to fob off an over-zealous suitor.

Hollywood Calling

Need some numbers to pad out your address book? Here are five useful 555 numbers for the next time you want to call one of your imaginary friends:

- *Ghostbusters* – If you have trouble with poltergeists, ghouls or restless spirits, the number to call is: 555-2368.
- *Bruce Almighty* – God's number came up on Jim Carey's pager as 555-0123, although only on the DVD version, after a mistake in the cinema release meant God actually had a working phone number.
- *The A-Team* – Hannibal Smith's mobile phone number was 555-6162 . . . 'If you have a problem and no one else can help, and if you can find them . . .' Oh wait! I think I've got Hannibal's number in my phone. Yup, there it is.

- *The Sopranos* – Tony Soprano's number is 555-0157. Just pray this number never flashes up on your screen as an incoming call.
- *The Simpsons* – 555-X-TERM-N-8 is the number that will bring . . . err . . . animal control to your door.

100

BUY LOW, SELL HIGH – THE FTSE 100

Despite business journalists getting much better at explaining what they're talking about when they mention 'the FTSE 100', many people still think of it as a mysterious financial monster that spells joy or gloom, dependent on what day it is.

But if you have a pension or other investments, there's a chance they will rise and fall based on the performance of this index, so it's worth having a basic understanding of what's going on.

This isn't only the case if you live in the UK, the home of the FTSE. If you invest elsewhere you have a similar interest in what's happening – because the world's stock markets have an unerring habit of following one another up and down.

A Beginner's Guide to the Stock Market

Firstly, let's try to clear up any confusion by turning to the FTSE website. This tells us: 'The FTSE 100 is a market-capitalisation weighted index representing the performance of the 100 largest UK listed blue-chip companies, which pass screening for size and liquidity.'

Well, that's as clear as mud, so let's start again.

- In the UK, shares in companies are traded on the London stock market.
- Traders buy and sell shares for their clients who are often large corporate institutions (like your pension fund) but also can be smaller private investors (like your average millionaire).
- The reason firms sell shares in the first place is to get access to cash and to expertise.
- The reason people buy shares is that they hope the value of them will go up and/or they will make money from the company paying out dividends.

If you own a company that has shares which can be traded by the public then, depending on how valuable investors think your firm is, you will be put in one of the various categories that fall within the 'FTSE Index Series'.

At the top end of the scale is the Daddy of the stock market – the FTSE 100. This index comprises the hundred most highly capitalised companies. You work out what a company's market capitalisation is by multiplying its share price by the total number of shares the firm has. (If you watch financial news you will regularly hear people talking about a company's 'market cap'.)

Some Key Facts About the FTSE 100

- The value of all the money invested in these companies' shares was £1.64 trillion at the end of 2011.
- It contains companies that cover everything from aerospace to travel and leisure, although it has a particularly large number of firms dealing with commodities, such as miners and oil explorers, at the moment.
- There are thousands of other businesses that also have shares you can buy and sell. The reason so much emphasis is put on these companies is that members of the FTSE 100 make up around 81 per cent of the total value of the UK market.
- Every three months the numbers are crunched and the FTSE 100 is updated to include any firms that merit promotion from the FTSE 250 (a wider index representing the next 250 biggest firms) and to remove any laggards, who face the ignominy of dropping out of the top flight.

The Big Bang

Sadly there is no 'open cry' system any more whereby people in funny jackets yell at each other. That went out in the 1980s when the system was automated – a development called the 'Big Bang'.

All the deals that are done are instantly fed into the indices, including the FTSE 100 index, which are updated by FTSE International, which is jointly owned by the London Stock Exchange and the *Financial Times* newspaper. This explains its original name: the Financial Times Stock Exchange.

It's All About Confidence

The ups and downs of the FTSE 100 reflect how much confidence investors have in those firms on any given day. A company's share price can be affected by:

- An announcement that it is doing well or badly
- News from its sector and its rivals
- Events in the UK or abroad that impact on it in some way

While the firms that make up the FTSE 100 are equal – in that they have all made it into the top flight – within the index some are more equal than others. Several companies have much bigger market caps than their compatriots and the larger they are, the more weight they carry when it comes to moving the market. So a change in the share price of one company can alter the index much more than if another, smaller firm, sees its shares go up and down. In fact the top 10 companies make up almost half of the FTSE 100 in terms of total weighting.

> **In early 2012 the top three firms were banking giant HSBC, whose shares made up 6.4 per cent of the index, oil firm BP with around 6 per cent, and telecomms group Vodafone, which came in at 5.7 per cent.**

We're All In It Together

Often people equate the performance of the British stock exchange with the health of the UK economy. The two are interwoven but if the FTSE 100 drops on any given day it might not have anything to do with Great Britain because so many foreign companies' shares are traded on our stock market. The London Stock Exchange is presently home to companies from around 70 different countries.

This means that a protest in Kazakhstan or a flood in Africa might cause your pension fund to drop in value. Then again, a company striking oil off the coast of Indonesia might suddenly put extra pennies in the retirement kitty.

It's globalisation before your very eyes.

A Brief History of the London Stock Exchange

1571: The Royal Exchange is founded, based on the Antwerp stock exchange, and opened by Elizabeth I.

1698: John Castaing opens Jonathan's Coffee House where he is the first person on record to publish lists of stock and commodity prices, which he calls 'The Course of the Exchange and Other Things'.

1698: Dealers are thrown out of the Royal Exchange for being too rude and mannerless. Instead they take their services into the nearby streets and coffee houses, with a particular favourite being Jonathan's Coffee House in Change Alley.

1761: Jonathan's becomes a central hub of trading after a group of 150 stockbrokers form a club there to buy and sell shares.

1773: The brokers outgrow the coffee house and instead build their own building in Sweeting's Alley. They have a dealing room on the ground floor but keep a coffee room upstairs. They soon start to call it 'the Stock Exchange'.

1854: The original Stock Exchange is rebuilt.

1973: Women are admitted as members to the market for the first time.

1986: The deregulation of the market takes place, known as the 'Big Bang'. Amongst other things, this meant that trading moved from being face-to-face on a market floor to a rather less sociable, but quieter, system of computers and telephones.

3:16

THE RAINBOW MAN AND 'JOHN 3:16'

You might have noticed a resurgence in signs bearing the message 'John 3:16' amongst spectators at sporting events recently, particularly American ones.

Readers who watched US sports in the 1980s may remember the origin of this phenomenon: a man called Rollen Stewart, known as the Rainbow Man for the multi-coloured wigs he wore as he held up his home-made 'John 3:16' signs.

Spreading the Word

Stewart was an ardent Christian and his signs were a reference to the verse from John's Gospel which reads: 'For God so loved the world, that he gave his only begotten Son, that whosoever believeth in him should not perish, but have everlasting life.'

Soon 'John 3:16' was seen at sporting events, music concerts and on clothes, as people enthusiastically adopted it, either because of their religious beliefs or just because they liked the crazy guy in the giant afro.

Fall from Grace

After Rollen's unfortunate arrest and life imprisonment in 1992 (he took a maid hostage and threatened to shoot down aeroplanes from his Los Angeles hotel room), the signs all but disappeared. Some ardent, and possibly unhinged, sports fans kept waving the placards but not as a religious statement, rather because they were equating their favourite star to the Son of God.

The reason the sign appears much more these days is down to an American football quarterback called Tim Tebow. In 2009, as a religious statement, he wrote 'John 3:16' in white lettering on his eye black (the grease that American footballers apply under the eye to reduce glare). This sparked renewed interest in the expression, no doubt both blasphemous and religious.

100 PROOF
HOW STRONG IS YOUR LIQUOR?

There's a chance that the strength of your favourite alcoholic tipple, particularly if it is from the USA, will be measured in two different ways. It will almost certainly have an 'alcohol by volume' (ABV) number, which measures the percentage of the drink that is alcohol. But it might also boast it is '100 Proof', or some other level thereof. It is often accompanied by a figure in degrees, for example 80° Proof.

This secondary measure is slightly archaic and is more a marketing tool than anything else these days, just because it sounds quite cool. Its origins go back as far as sixteenth-century Britain when proof meant 'test' – as in 'the proof of the pudding is in the eating'.

Passing the Proof

Proofing was important in booze terms because it was an

effective way of assessing how strong a drink was. It was known that if you doused gunpowder with a sample of the drink in question and then tried to light it, the sodden gunpowder would not ignite unless the liquor on it was at least 57.15 per cent ethanol.

This had two practical applications. Firstly, it allowed extra taxes to be levied against the alcoholic drink because it was known to be above a certain strength. Secondly, the British Navy gave sailors rum as part of their pay and conditions (something it started in the West Indies, after the Navy found beer did not keep in the warm climate). By 1731 'the rum ration' was in general use with each man being issued ¼ pint twice a day, which was drunk neat. The sailors would use the gunpowder test to make sure their ration was not being watered down.

If the gunpowder did light, then those doing the experiment, be they prying taxmen or suspicious sailors, knew that the drink had passed the test – or the proof. It was therefore termed '100 degrees proof'.

Relative Values

The routine was simplified in the United States in the mid-1800s when it was decided that the British system of $^4/_7$s (which is around 57 per cent) was too complicated. Instead they declared that 50 per cent alcohol was now to be equivalent to 100 Proof, as this level of ethanol was deemed 'typical' of strong distilled liquor.

All of which means if you buy an American spirit which is 100 Proof and a British one that makes the same claim, the UK drink will be considerably stronger, since 50 per cent ABV is just 87.5 degrees Proof in British terms. Having said that, if you're concerned at your booze not being strong enough at 50 per cent, then you've probably got bigger problems to worry about.

FAC 191
THE VINYL
COUNTDOWN

Codes like this can be spotted on LP and CD album covers, although the advent of digital music makes them rather more difficult to find these days. There's a very small chance you might come across them on a coffin or even a lawsuit, for reasons that will be explained shortly.

FAC 191 is an example of a record label catalogue number, which is a code usually assigned to an album or single that reveals all sorts of details about it. Each time a record label releases a new album or single it assigns it an identification number – this is the catalogue number. It is used for tracking purposes by both the label and the distributor.

But it can be much more than that; it can be a fascinating, revealing number, particularly if you are a collector, music junkie or a professional like a radio producer. This is because it tells a colourful tale about the music that you hold in your hand.

Music by Numbers

Once upon a time, catalogue numbers were typically printed on the spine of a CD box or on the back of record sleeves. You might also find them on a CD itself or on the info label stuck on a record. Sometimes you'll find them in other places, such as incorporated into the artwork.

There aren't any rules as to how a record label chooses its catalogue numbers, but they typically include both numbers and letters – often some portion of the record label's name, combined with figures that signify the number of the release for that company.

In the very early days – 1908 to be precise – Columbia Records started a series of 78 rpm records with an 'A' prefix and four digits, going from A0001 to A4001 (which was reached in 1923). A different series, beginning with A5000 in 1908, was used for 12-inch 78s. All record companies have a variation on this original theme.

In a more recent example, Virgin Records kicked off its releases in 1974 with VS101 for Mike Oldfield's first single. As it released other formats it added prefixes, so a 12-inch single got VST and EPs received VSEP. The more formats in use, the longer the prefixes would become. The 1993 UK number one 'I'd Do Anything For Love (But I Won't Do That)' by Meat Loaf had the catalogue number VSCDT1443.

Cracking the Code

There are a number of very practical pointers that these catalogue numbers offer, particularly to the collector:

- In older vinyl records you can find numbers indicating the speed this record is played on. This was designed to help people to buy the correct record, as there were three different types on the go when they were originally released.
- The year of a release is often encrypted within the catalogue number, separating an original from often less valuable re-releases.
- The country of origin might also be specified and, again, particularly in terms of vinyl, this could make a big difference to its value and whether a trip to the auction house might be worth your time.

So if you're after a Beatles original release, check the record company's catalogue for the number assigned to that particular release to make sure the code matches the one in your hand. Then again, a re-release might be just what you're after so you can get all the bonus tracks, bigger booklet and other interesting paraphernalia.

Factory Facts

At the end of the day there are no rules. As long as the numbers help the label and distributor track releases, anything goes. Just to prove this point, take Factory Records, the independent Manchester-based label, which took extreme liberties with its cataloguing. It had some standard numbering rules, such as the last digit of the code, which is likely to designate the following:

1 – Factory Corporate
2 – Happy Mondays (singles)
3 – Joy Division / New Order (singles)
4 – Durutti Column
6 – Factory Classical

But this being Factory Records, these rules are subject to numerous exceptions. The company also had special numbers that were reserved for significant Factory output; for example FACT 25 for Joy Division's *Closer*, FACT 50 for New Order's *Movement* and FACT 75 for New Order's *Power, Corruption & Lies*.

In addition, many of its catalogue numbers had nothing to do with traditional releases. Its very first catalogue number, FAC 1, was an event. FAC 7 was the number attributed to some stationery, while FAC 61 was a lawsuit.

FAC 501 was on Factory co-founder Anthony H. Wilson's coffin.

This just leaves us with the question: 'What was FAC 191, the number chosen out of all the others to be the title of this chapter?'

That was the number given to a particularly important character at Factory's Hacienda night-club: its cat.

3.142
BECAUSE EVERYONE LIKES A BIT OF PI

Pi, as every school child knows and almost every adult has forgotten, is the magic number that helps tell you how big the circumference of a circle is. (Remember the formula $C = 2\pi r$?) It is usually stated as 3.142 but the number goes on infinitely.

Pi has been used for millennia, particularly for building projects, by civilisations such as the Ancient Babylonians, who narrowed it down to 3.125, and the Ancient Egyptians, who used 3.160.

The modern formula for pi was created in 1706 by mathematician John Machin, who, if you want to get really technical about it, came up with this handy calculation:

$$\frac{\pi}{4} = 4\arctan\left(\frac{1}{5}\right) - \arctan\left(\frac{1}{239}\right)$$

We're Going to Need a Very Large Gravestone ...

Pi has inspired people to do some pretty crazy things, not least those who lived before the advent of the computer. Seventeenth-century German mathematician Ludolph van Ceulen dedicated most of his life working pi out to 35 decimal places, for example. The solution he came up with is inscribed on his gravestone, just to make sure he hadn't wasted the best part of his life for people to go and forget the number.

Van Ceulen would no doubt be impressed, and also quite upset, that by 2010 a computer scientist, Fabrice Bellard, had computed pi to nearly 2.7 trillion digits. It took him 131 days.

An Astonishing Feat of Memory – and Bladder Endurance

The official Guinness World Record for memorising pi is held by China's Chao Lu, who recited pi from memory to 67,890 digits on 20 November 2005. It took him 24 hours and four minutes. According to rules set by the *Guinness Book of World Records*, the time taken between his stating two numbers could be no more than 15 seconds, so he could take no breaks for food or to go to the toilet during the recitation. Sadly, his error on the 67,891st digit was saying it was a '5', when – as of course you know – it was actually a '0'. Mr Lu had been practising it for four years before the attempt and had hoped to reach 91,300 digits.

4.7 BILLION
WINNING THE
RATINGS WAR

Who wins the 'ratings war' is an absolutely crucial measure for TV channels, programme-makers and advertisers.

Ratings determine how effective advertising is and how much TV channels can charge for it. They will often decide which shows survive and prosper, while condemning others to obscurity after one series, or even earlier.

You might have lavish sets and the finest Shakespearean actors, but if the ratings go against you then it's usually game over.

That might sound unfair, but if you're an advertiser spending vast sums of money, you want to know lots of people are going to see your ad.

> The most-watched TV show in the UK, excluding sports and
> news coverage, was first aired on Christmas Day 1986. As
> everyone who is old enough to remember it will know, this
> was the day that Dirty Den served his divorce papers on Angie
> in *EastEnders*. Over 30.1 million viewers tuned in to witness
> the historic event.

Learning the Lingo

When people refer to successful programmes or advertising
campaigns they often talk about how many 'ratings points'
they got. Each of these points corresponds with 1 per cent
of the potential audience. So, for example, if 25 per cent of
all targeted televisions are tuned to a show that contains
your ad, you have 25 rating points.

Each time the relevant show is on air you collect more
points and these are added together to get the total number.
This is usually then compared against a benchmark to work
out whether the whole thing has been a success for the
programme-makers, the networks and the advertisers.

Rapid Results

You might have noticed how quickly newspapers and
websites are able to report how many million people have
watched a particular programme and how this compared
with the rival on the other channel. Details are collected on
a minute-by-minute basis, something that appears to be a
feat Santa Claus and his chimney-hopping reindeer would
be proud of.

Just how on earth do they get so much data on so many people, and in such a short time? The simple answer is those responsible for ratings don't deal with anywhere near the number of people who actually watch the programmes, so Santa can rest easy.

Who's Watching You Watching TV?

Different bodies are responsible for collecting this information in different countries. In the UK it is the Broadcasters' Audience Research Board, or BARB, that is in charge of tracking who watches what, while it is OzTam in Australia. In the USA and Canada, for example, Nielsen Media Research holds sway.

However, at their heart, the collection methods are very similar. The chief way of doing it is called 'statistical sampling', where the companies use a 'panel' of television-owning households, whom they claim represent the viewing habits of the country as a whole.

In the UK this means 5,100 private households are selected to represent the nation. (Not quite as good as lining up with the national sports team, but hey, everyone's got to play to their strengths.) The reasons they are chosen are based on all sorts of complicated metrics – or in industry-speak, they are selected by 'a multi-stage, stratified and unclustered sample design'. Then the following steps are taken to hook them up to the database:

- Firstly, the household has to agree to join the panel – which is a relief, because otherwise the whole thing would be rather Orwellian.

- Each TV in a home, as well as personal video recorders and VCRs, is connected to a small box. This meter automatically identifies and collects information about channels that the panel member is watching.
- To make the system accurate all panel household residents and their guests need to register their presence when they are in a room that has a television set switched on. Each individual does this by pressing a button allocated to them on the 'peoplemeter' handset. Each push of the button is recorded so the meter knows just who is watching what and when. This is important as it will point out what type of person watches a show, be they young, old, male or female.
- Throughout the day the meter system stores all the data of what you are viewing and while you slumber the meter goes to work, with the data downloaded each night between 2 a.m. and 6 a.m.

This account is a very simplified version of what happens, as monitoring firms also carry out lots of surveys and other research to make sure they're getting the right information.

Digital Killed the TV Star

Companies that do this kind of measurement have found it has got much more difficult in the last few years. They have had to make all sorts of changes because lots of people now watch TV via 'play-back' services. (The industry calls this viewing 'timeshift', which sounds like a film title ready-made for the likes of Jean-Claude Van Damme.) Furthermore, many people ignore TV altogether, choosing to watch their favourite shows on their computer or mobile phone.

This has led to accusations that certain programmes have been unfairly cancelled in the past when they were well watched, but just not on TV. Critics say that if you have a technologically sophisticated audience, who are often younger, programmes could unfairly suffer in the ratings simply because tracking has not kept pace with people's viewing habits. Such critics cite booming DVD or download sales of a programme that appeared to fail on TV as proof that the system is flawed.

The firms reply that they are now using ever-more sophisticated methods, which offer accuracy across multiple platforms, particularly in the case of website performance. They also remind people that while technology marches on, TV still remains the favourite viewing method.

For example, Nielsen's research shows that in the USA, the couch potato capital of the world, almost one in three TV households – 35.9 million – owns four or more televisions. Around 290 million Americans own at least one TV, in contrast with 211 million Americans who are online and 116 million (ages 13+) with access to the mobile web.

The Big Day

Just to show quite how accurate TV ratings can be, let's look at 2011's Royal Wedding of HRH Prince William and Catherine Middleton (now the Duke and Duchess of Cambridge), which took place on 29 April 2011. According to BARB:

The ceremony, which ran from approximately 11 a.m. to 12.10 p.m. achieved an average audience of 26 million UK viewers. Its peak of 26.3m viewers was at 11.08 a.m., when the Duchess of Cambridge reached the altar. There was another spike at 12.10 when the Duke and Duchess appeared on the steps outside of Westminster Abbey, after which viewing numbers declined. At 13.30 the numbers peaked again, almost hitting the 25m mark when they emerged on the balcony of Buckingham Palace and had the long-awaited kiss.

> **It is estimated that 2 billion people in more than 180 countries around the world saw reports, photos and TV pictures of the Royal Wedding.**

The Biggest Viewing Figures *Ever*

According to the US number-crunchers Nielsen, the 2008 Beijing Olympic Games attracted the largest global TV audience ever. Between 8 and 24 August, 4.7 billion viewers – or 70 per cent of the world's population – tuned in to watch the Games, it said. By comparison, 3.9 billion watched the 2004 Athens Games, while 3.6 billion followed the 2000 Sydney Games on TV.

Host nation China led the viewing with 94 per cent of Chinese viewers tuning in to the Olympics TV coverage, Nielsen reported. South Korea, though a much less populous nation, also recorded 94 per cent audience reach. Mexico followed closely, with 93 per cent of all viewers in that country following the Olympics on TV.

In the USA, the Summer Games ranked as the most viewed TV event ever, with a total audience of 211 million and an average daily audience of 27 million people.

It's well documented that the Super Bowl in the US can command ridiculously high prices for advertising during the game. In 2012 the average 30-second spot during the Super Bowl cost around $3.5 million, the highest price in history – that's up 84 per cent from ten years ago.

It's a pretty safe bet for advertisers willing to splash the cash that they'll get value for money in terms of audience numbers. In 2012, the Super Bowl was crowned the most-watched (one-off) TV broadcast ever, for the third year in a row. An estimated 111.3 million people watched the New York Giants defeat the New England Patriots, 21–17. All of which makes the adverts a remarkably reasonable price of about three cents per viewer.

40–0

GAME, SET AND MATCH

There are few games that have more a bizarre system of scoring than tennis. You start with 'love', then progress through 15, 30 and 40, perhaps hitting 'deuce' and 'advantage' along the way. Only then can one player win a game. But you still have to win a certain number of 'sets' before you can win the overall match. There seems to be no rhyme or reason to it whatsoever.

The origins of tennis lie in France where it was popularised as a game of the aristocracy from the sixteenth century onwards. It started as 'Jeu de Paumme' – the game of the palm – before rackets became standard. It is thought the name 'tennis' has its origins in the fact that French players would shout 'Tenez!' at the beginning of each game, which meant something like 'ready' or 'play'.

Sixty Minutes to Victory

Unfortunately there are no definitive answers as to why such a strange scoreline exists, but there are plenty of theories, some more believable than others.

Let's start with the potential scores 15, 30 and 40. One leading theory suggests that this sequence comes from the face of a clock, which would always be present by old tennis courts. Each time a point was scored the hands would be moved one quarter of the way round – i.e. to 15 minutes past, then 30, then 45. When one hand did a full circle to 60 then a game was over. This doesn't explain 40, but proponents of this theory say it is simply an abbreviation of 45.

A second theory also revolves around the number 60 signifying the completion of a game, but has nothing to do with a clock. This idea says that since the Middle Ages in France, 60 was a 'complete' number, much in the way 100 is considered a round number today. (This might explain why 70 in French is 'soixante-dix' or 'sixty and ten'.) So it followed that 15, 30, and 45 were handy divisions of 60. Again, it is said that 45 was then cut down to 40.

Cannon Fodder

There is a third theory, which is altogether more exciting, if considerably less likely. This one goes that the scoring system was based on the different gun calibres of British naval ships. When firing a salute, the ships first fired their 15-pounder guns on the main deck, followed by the 30-pounder guns of the middle deck, and finally by the 40-pounder lower gun deck.

Indeed, tennis as we know it was invented by a military man, Major Walter Wingfield, in the late nineteenth century. He created 'lawn tennis' in 1874, which was quickly shortened to tennis. (This was much to the anger of players of the original game, who snootily started calling their version 'real tennis'.)

The big problem with the gun-calibre theory is the scoring does not correlate with the actual size of cannons that were in regular use. If they did we'd have even more confusing and random scores such as 12, 32 and 42 (numbers that refer to the weight of the shot, in pounds, which the cannons fired). So much for the cannon theory.

The Mystery of Love

Now, let's put the guns down then and start thinking about love. Why is it that tennis begins with both players on 'love', rather than zero, as in most other sensible games? Again, there are three mainstream theories and you can take your pick.

The first is simply that competitors are playing for the love of the game. A variant on this suggests that at the beginning of each match the players still 'love' each other as they are yet to score any points.

Another idea is that 'love' is a corruption of the French words for 'the egg', *l'oeuf*, because an egg looks a bit like a zero. Tenuous? Very possibly, although the same reasoning is applied to 'being out for a duck' in cricket. Again this means getting no points, with the 'duck' being an abbreviation of 'duck's egg'.

A third theory says that the term comes from the Dutch phrase *'iets voor lof doen'*, meaning 'to do something for praise'. So, even if you get hammered by your opponent and score no points, you still get the praise (or 'honour' in some translations) for taking part.

If you tweak the phrase slightly you get *'iets voor lief doen'*, which means 'to do something for love', taking us back to the first explanation about taking part for the fun of it.

Incidentally, translating this sentence on some language websites gives you the phrase 'do anything for love'. This might finally clear up what Meat Loaf was talking about in his song 'I'd Do Anything for Love (But I Won't Do That)': it's a coded reference to a dislike of racket sports.

I'll Be Deuced

Finally then, we are left with 'deuce' in our menagerie of strange tennis terms. This one is more easily explained than the others, or at least more rationally explained, because we still can't be sure of its origins. The word seems likely to have come from the French *'a deux du jeu'* – two points from the end of a game. The English, no doubt to the disgust of the French, then corrupted it to deuce, where it has stayed.

The Tie-Break

One of the few scoring rules in tennis we can explain with certainty is the tie-break, because it was invented relatively recently. To prevent long matches, a tie-break is played to decide the winner of a set when the score is six games each. The first player to score seven points, with a minimum two-point lead, wins the tie-break and the set.

James 'Jimmy' Van Alen, a former singles and doubles champion, invented the tiebreaker (as it was originally known). It was first used in a nine-point 'sudden death' format at the 1970 US Open.

One of the great catalysts for the acceptance and uptake of the tie-break was the 1969 match at Wimbledon between Pancho Gonzales and Charlie Pasarell. It lasted five hours, 12 minutes and included 112 games. This was made all the more remarkable by the fact that players did not sit down between games in those days, merely stopping for a quick drink by the net if they wanted it.

16

DIGITS ON A
CREDIT CARD

Take a moment to pull the credit card out of your wallet or purse and have a look at the number on the front. Most people assume it's a randomly generated number, the length of which can be attributed to the enormous volume of cards in circulation.

These assumptions are wrong on both counts. This number is about as far from random as lenders can make it and contains a considerable amount of information. So let's break it down starting with the first digit, reading from left to right.

All About MII

The first number is called the Major Industry Identifier, or MII amongst lenders. This identifies the category in which your credit card company sits. The system goes like this:

1 and 2 = airlines
3 = travel and entertainment
4 and 5 = banking and financial
6 = merchandising and banking
7 = petroleum
8 = telecommunications
9 = national assignment

This number is then combined with the following five digits to make up the Company Identification Number, which denotes exactly which business approved the credit card. This system allows for the total number of issuers to reach one million. You'd hope we never need that many, but then people love nothing more than borrowing money to pay for things they can't afford, so it's probably only a matter of time.

> **Diner's Club, which was introduced in 1950, has the honour of being the original credit card issuer. The original was made of cardboard and was first used by its creator Frank McNamara at Major's Cabin Grill in New York – an event the company likes to call 'The First Supper'.**

Card Codes

Continuing along the face of your credit card, the digits from the seventh number right up to the penultimate one denote your account number. Some companies use up to 12 digits here, giving them a possible one trillion combinations.

However, most stick with nine, restricting them to a mere one billion account numbers per issuer.

Here is a list of some of the better known issuer codes as well as how many digits in total their cards should display:

Issuer	Identifier	Card Number Length
Diner's Club	300xxx–305xxx, 36xxxx, 38xxxx	14
American Express	34xxxx, 37xxxx	15
VISA	4xxxxx	13, 16
MasterCard	51xxxx–55xxxx	16

How to Check the Checksum

The final number on the card is the one that will test your interest in numbers to its limit. This is the 'check digit', or 'checksum', which validates your credit card through something called the Luhn algorithm.

Credit cards are only given the thumbs-up in the first place if they pass this test. The Luhn check is also used to verify existing card numbers. In either case, if a credit card number does not satisfy this check then it is not valid.

Now concentrate, because this is how it works:

Step 1
Take your card number and double the value of every second digit, beginning from the right.

To use a common example, for a credit card number 4417 1234 5678 9113, reading from the right and doubling every other number, we get:

(4 x 2) [4] (1 x 2) [7] (1 x 2) [2] (3 x 2) [4] (5 x 2) [6] (7 x 2) [8] (9 x 2) [1] (1 x 2) [3]

Let's ignore the numbers in square brackets we jumped over for a minute. Just looking at the numbers we created by doubling every second digit, we are left with:

8 2 2 6 10 14 18 2

Step 2
If doubling a number results in a two-digit number, add up the digits to get a single digit number.

So in this case 10 = 1 + 0, 14 = 1 + 4, 18 = 1 + 8, leaving us with:

8 2 2 6 1 5 9 2

Step 3
Now we need to bring back in the original numbers that we'd ignored and add everything up. For ease, you can split the numbers in the way they are broken up in groups on the front of your card.

In our example this creates the sum of: (8+4+2+7) + (2+2+6+4) + (1+6+5+8) + (9+1+2+3). Adding this all together gives us a total value of 70.

So why have we gone through this rigmarole? Well, the key is the total number you come up with has to be divisible by 10. If, for example, the final number on our fictional card had been a 4 rather than a three, then we would have finished up with 71, which would demonstrate that it is not a valid card number.

Using this technique anyone can create what would be deemed a 'valid' credit card number. However, wannabe fraudsters should take note that it won't create a working credit card number, so you're going to have to work harder than that at your financial skullduggery.

Safety in Numbers

Finally, the three-digit security code, which you should be asked for when making a purchase online or over the phone, is an extra layer of security to prove you are who you say you are.

The three digits, usually found on the back of the card, are known by different names including CVV Number, 'Card Verification Value', and CSC numbers, 'Card Security Code'. (Just to be special, American Express puts its numbers on the front and goes for four instead of three.)

The concept is quite simple: the security code is only ever printed on the card. It is not contained in the magnetic stripe information, nor does it appear on any sales receipts or statements. This provides assurance to the vendor that you have the card in your hands when you are buying.

101
THE ROOM FROM HELL

This is a place that you never want to end up in: it is the name of a torture chamber in George Orwell's terrifying vision of the future, *1984*. Room 101 is situated within the 'Ministry of Love', one of the departments of the despotic government that runs the country of Oceania, in which the novel is set.

The room is used as the final means to break the spirit of prisoners by subjecting them to their worst nightmare. The place is summed up in this quote from the book:

'You asked me once,' said O'Brien, 'what was in Room 101. I told you that you knew the answer already. Everyone knows it. The thing that is in Room 101 is the worst thing in the world.'

Not nice. Not nice at all.

The inspiration for the name is, fortunately, a little more amusing. It is said that while working for the BBC between 1941 and 1943 George Orwell had to attend some very boring committee meetings. These meetings, which taxed his patience so much, took place at 55 Portland Place, London . . . in Room 101.

SPF 15
BURN, BABY, BURN

Are you one of the many folk who believe a sunscreen marked SPF 30 will keep out twice the amount of harmful ultraviolet light as SPF 15 sunscreen? If you are then you're wrong – and for health reasons it's probably in your interest to read on.

The One Travelling Companion You Really Need

These days most people are pretty good at putting on sunscreen because they know that the temporary glow they foster during their holiday break could actually result in something considerably nastier in the long term.

In fact, if exotic sun-drenched locations advertised properly then the golden beaches, tropical lagoons and pearlescent seas should probably be shown alongside photos of deepened wrinkles, leathery skin and – the ultimate holiday accessory – skin cancer.

But if spending your whole life without leaving your own rain-sodden nation is too depressing to contemplate, you should thank your lucky stars we have sunscreen. Then you should actually take a moment to work out what sunscreen you actually need to put on.

Extreme Protection

People tend to fall into one of two camps. The first are those who choose sunscreen that has a very high sun protection factor (SPF) because they are afraid of the consequences of exposure. At the other end of the scale are the people who choose a very low SPF because you just can't beat looking like a lobster and being unable to put on your clothes due to the searing pain of sunburn. In fact there is also a third camp: the people who combine low- and high-level sunscreen in the hope that adding SPF 15 to SPF 30 will create SPF 45. It does not work. Stop it, you're just wasting the stuff.

So How Does It Work?

Here's the science: sunscreens are chemical agents that help prevent the sun's ultraviolet (UV) radiation from reaching the skin. There are two types of ultraviolet radiation we're concerned about: UVA and UVB rays.

- We can't see these rays like we can see other light because they're so unbelievably small.
- UVA is the longest, but its wavelengths only come in at 320–400 nanometres, so we're talking in terms of billionths of a metre.

- UVB is the one that gives us sunburn and is the chief cause of skin cancers.
- UVA rays penetrate the skin more deeply and are associated with signs of ageing such as wrinkles, leathery skin and saggy bits.
- However, increasingly scientists believe UVA also teams up with UVB to cause cancer.

So, to give you peace of mind, those self-same scientists developed sunscreen, or sunblock, which does one of two things. Either it works by using an organic chemical compound (such as oxybenzone) to absorb the ultraviolet light, or it contains an opaque material (such as titanium dioxide or zinc oxide) that reflects the light. Some have a combination of both.

Length Not Strength

As we well know, sunscreens are graded by SPF, a measure of the level of protection you are getting against UVB rays – but not UVA. Crucially, it is not a measure of the strength of the sunscreen. Rather, it tells you the length of protection it offers.

Experts recommend you use an SPF of at least 15. Using this level of protection as an example, if it takes 20 minutes for your unprotected skin to start turning red, using an SPF 15 sunscreen theoretically prevents reddening for 15 times longer – that's about five hours.

But beware: the intensity of the sun is dependent on where you are and what the weather conditions are like, and it will change throughout the day. Many people also forget ultraviolet radiation is reflected off sand, water and ice, again intensifying exposure. That's why it is recommended you reapply sunscreen every two hours.

> Studies have shown sunbathers usually only put on between one quarter and a half of the amount manufacturers use in their tests. So you could be effectively making that SPF 30 into SPF 8 by spreading it too thinly. One way to make sure you're putting enough on is take the shot glass you somehow ended up with after last night's holiday shenanigans and fill it with sunscreen. This is about one ounce, the recommended amount for each application.

Stopping Ultraviolet in Its Tracks

Let's dispel another myth about sunscreen: SPF 30 does not block double the amount of rays that SPF 15 does. According to the New York-based Skin Cancer Foundation, SPF 15 blocks approximately 93 per cent of all incoming UVB rays, SPF 30 blocks 97 per cent, and SPF 50 blocks 98 per cent.

At the other end of the scale, the Australian government puts the protection offered by SPF 4 at 75 per cent, with SPF 8 blocking 87 per cent of rays. In a country where it is estimated two out of three of the population will develop some form of skin cancer during their lifetime, the government unsurprisingly recommends factor 30 for everyone. Why not higher? In the words of one Australian doctor, 'so people don't think they're wearing a suit of armour'.

The simple fact is no sunscreen can block all the rays, so take a hat.

You can now get SPF in factors of 90 or 100, which will make minimal difference to the UV that hits your skin in percentage terms. Some dermatologists scoff at these ultra-high SPFs and dismiss them as marketing. Others say they could be helpful if you wear them over many years because the amount of UV rays they block will all add up in the end.

A Final Word of Warning

Since SPF is all about UVB rays, it is also important to ensure your sunscreen has UVA protection in it. UVA doesn't burn the skin and without protection there could be all sorts of damage being done without you even knowing it. The common phrase you're looking for on a bottle that fights UVA and UVB is 'broad spectrum protection'.

Now all that remains is to wish you a happy holiday . . .

1
PARTRIDGE IN A PEAR TREE

One thing that is impossible to forget over the festive period is the song 'The 12 Days of Christmas'. It's a surreal tale of birds sitting in trees and the aristocracy living it up – all set to a dozen drummers making one heck of a noise.

As you'd expect from such a strange set of lyrics there are a number of explanations as to their origin. There's no question that at some point it was written to celebrate the official Christmas season. But why it was written is a matter of some fierce argument.

Secret Song

Many Catholics believe it came into being in sixteenth-century England, after King Henry VIII created his own religion because the Catholic Pope wouldn't let him divorce his wife. This was the beginning of a period of horrific persecution for the Catholics, who had to worship in secrecy or face punishment.

With this is mind, the theory goes that the song was written as a way to teach Catholic doctrine, but camouflaged as a children's song. In this version of events, 'my true love', who gives all the gifts, refers to God, while 'me' is the individual Catholic. From then on each number and corresponding event has an important meaning. For example, the 'the ten lords a-leaping' are the Ten Commandments, while the 'four calling birds' are the four books of the New Testament.

A Holy Theory

But there are a number of issues with taking this explanation as gospel, so to speak. Firstly, no one seems to be able to offer any historical proof it was written for this reason. Secondly, the song seems to cover shared tenets of Christianity, meaning you could sing it all you like as a Catholic without any raised eyebrows. If instead of pipers there had been '11 popes a-poping' then there would indeed have been trouble.

Other questions are asked about how tenuous some of the analogies seem (but then you'd want them to be if you stood a chance of being burned at the stake, one supposes), as well as what songs were there to teach kids Catholic doctrine for the rest of the year?

A Week of Wings

The song has a number of variations, which offer different birds and/or weird gifts on certain days. There have been bells ringing, cocks a-crowing and both hares and hounds running, to name but a few.

Incidentally 'four calling birds' is actually a recent mispronunciation; it should be 'four colly birds', or blackbirds. It's also said 'five gold rings' refers to the ring-necked pheasant. If this is right then it means we've got birds from day one right through to seven when the maids take over with their milking.

> The earliest printed version of the song is in the 1780 children's book *Mirth Without Mischief*, where it was described as a 'memory and forfeit game' akin to many games we know today. The first person would start with one and the lyrics would build until someone got it wrong and had to do a penalty.

To be honest, no one can definitively 'prove' the origins of this song and it is one of those examples where it's pretty harmless to let people believe whatever they like. Then we can settle down to a good old wassail (see page 54) and celebrate happiness and goodwill to all.

The Secret Code

The Catholic 'secret meaning' to 'The 12 Days of Christmas' in full:

1. The partridge is Jesus and the pear tree the cross.

2. The two turtledoves are the Old and New Testaments.

3. Three French hens stand for faith, hope and love. Or the three gifts of the Wise Men.

4. The four calling birds are the four Gospels.

5. The five gold rings recall the Hebrew Torah (Law), or the Pentateuch, the first five books of the Old Testament.

6. The six geese a-laying stand for the six days of creation.

7. The seven swans a-swimming represent the seven gifts of the Holy Spirit.

8. The eight maids a-milking are the eight Beatitudes.

9. Nine ladies dancing are the nine fruits of the Holy Spirit.

10. The ten lords a-leaping are the Ten Commandments.

11. Eleven pipers piping represent the eleven faithful Apostles.

12. Twelve drummers drumming symbolise the twelve points of doctrine in the Apostles' Creed.

4.3 BILLION
IP ADDRESSES

Your Internet Protocol address (to give it its full name) is your computer's version of your mailing address or telephone number. In fact any device you have that connects to the Internet needs one of these.

These addresses are unique and allow other computers on the Internet to pinpoint your device and sort it from the billions of others connected to the network. Without IP addresses the Internet super-highway would just be one big messy traffic jam, with everyone honking their horns and shouting, and with no one paying the slightest bit of attention to what was being said.

When 4.3 Billion Is Not Enough

IP addresses are constructed using four numbers, each of which has up to three digits ranging from 0 to 255. A single dot separates the numbers. So you might see a number like this: 62.115.7.156. (This is actually not how the computer sees it because it works in binary, but it's good enough for our puny human brains.) This means the total number of possible IP addresses in the world is roughly 4.3 billion.

This might seem like a lot but they are already running out, a phenomenon known as 'address exhaustion'. Hundreds of millions of people continue to get online, and are doing so in multiple different ways. In February 2011, the Internet Corporation for Assigned Names and Numbers (ICANN) gave out the last of its unused large blocks of addresses under the current system, IP version 4.

Quite a Lot

So, the current system is being replaced by IP version 6. This should satisfy the doom-mongers since the new version will offer over 340 undecillion addresses. That's 340 trillion, trillion, trillion. Just to highlight what an unbelievably, unimaginably large number this is, here it is in numerical form:

340,282,366,920,938,463,463,374,607,431,770,000,000

The boffins think this is essentially an inexhaustible number of addresses. It also wins a prize for the biggest number in the book.

No. 5
The Scent of Success

Chanel No. 5 is probably the most famous perfume on the planet. It contains 85 different ingredients and is sold in bottles that are designed to look like whisky decanters.

It was launched in 1921, having been created by the perfumer Ernest Beaux for fashion icon Coco Chanel. The designer wanted to make something special for her best clients and when Beaux returned to her with ten samples for her approval, it was number five that was chosen.

A Fortunate Error

There is a legend that No. 5 stood out because of a laboratory mistake that created a smell very different to the other samples. Some say that having chosen it, Chanel just stuck with the number that was on the bottle, but others believe the choice of number five carried much more significance than that.

She reportedly told Beaux: 'I present my dress collections on the fifth of May, the fifth month of the year, and so we will let this sample number five keep the name it has already; it will bring good luck.'

The Magic Number

The creator of the 'little black dress' was certainly very superstitious, but the origin of those beliefs, and hence the inspiration for the perfume's name, are disputed. It is said the root of her superstition – and her faith in the number five – comes from her childhood.

While Coco Chanel became a member of the social elite as a world-famous fashion designer, her beginnings were much more humble. After her mother died when she was twelve years old, Gabrielle Chanel (as she was then) was packed off to a convent orphanage called Aubazine by her father, an itinerant market trader.

She spent her teenage years at the orphanage and there is evidence that this is where the number five became so important. Aubazine was a place filled with magical signs and symbols, including the number five, which was seen by the Catholic Cistercian order as a number representing purity.

Visitors to the convent tell of paths to its cathedral being laid out in patterns that repeat the number five, while the hillsides around it were said to be covered with the five-petalled rose, the Cistus.

The Secret of Its Success

How much her early life experiences impacted on her decision to call the world-beating perfume 'No. 5' is difficult to say. Chanel had a habit of surrounding herself with myths and legends, many of her own creation.

What can't be contested is its success; it is said one bottle is sold every 30 seconds, netting the company $100m per annum. Its continued allure after more than 90 years was perhaps best summed up by Marilyn Monroe in 1954. When asked about her nocturnal habits, she famously replied: 'What do I wear in bed? Why, Chanel No. 5, of course.'

8

PIECES OF EIGHT!
PIECES OF EIGHT!

Since Robert Louis Stevenson's book *Treasure Island* was published in 1883, most schoolchildren have been aware of 'pieces of eight' and their allure to swarthy, cutlass-swinging pirates. This was largely courtesy of Long John Silver's parrot, Captain Flint, who was trained to cry out: 'Pieces of eight!'

There's a good chance that parrot knew a good deal more about what pieces of eight actually were than you do. But Captain Flint was right to recognise the importance of these mysterious items, since they were actually the world's first global currency and a big draw for money-grabbing pirate types.

Cold, Hard Cash

The silver coins represented eight of the Spanish currency, the 'real', hence the name. The coin's full name is *peso de ocho reales* – eight pieces of reales.

In 1600, one of them would have been worth the equivalent of a modern £50 note, according to the British Museum.

They became the most common form of cash in the world from the late sixteenth century onwards and were an integral part of the global economy.

Here's the History

'You cannot write a history – especially an economic history – of the early modern world without engaging with pieces of eight,' according to the British Museum's curator Barrie Cook.

The reason for this is that at that time, Spain was building a vast empire that reached from South and Central America to the Philippines. The Spanish discovered large amounts of silver in Mexico and Bolivia, and pieces of eight became a major way of putting it to good use in funding their imperial ambitions.

> The coins carried different inscriptions but a common one was the moniker 'King of the Spains and the Indies' – referring to the Spanish monarchy and its global reach – as well as the coat of arms of the Hapsburgs, the ruling family. Many of the original coins were not round. More important were their weight and purity, and the Spanish words and pictures stamped on them to prove their authenticity. This often led to them being chopped up into eight pieces and used as small change.

The Human Cost

While pirate stories, right up to the recent upsurge in interest driven by the *Pirates of the Caribbean* film series, give 'pieces of eight' a certain swashbuckling romanticism, there is a dark history behind their creation.

Much of the precious metal used in the coins was mined from the 'silver mountain' at Potosi in Bolivia. The silver was first discovered in 1544 and led to a boom in the area, with the city of Potosi becoming one of the richest in the world.

But this came at a terrible cost. The indigenous population was decimated as the Spanish forced them to work in terrible conditions in the mines. When there wasn't enough manpower left amongst the locals, the Spanish brought in African slaves, who also died in their thousands.

> The volume of silver the Spanish extracted from the Americas was extraordinary. It rose from 148 kg in the 1520s, to 300,000 kg in the 1550s, and reached nearly 3 million kg by the 1590s.

The Spanish Dollar

Pieces of eight remained a dominant force in world currency for a long time. They were widely available in North America well into the nineteenth century and remained legal tender in the USA right up until 1857. In fact, the American dollar is even named after the Spanish coins: 'dollar' is an Anglicised version of 'thaler', a term originally applied to coins minted in the early sixteenth century in Bohemia, in the modern-day Czech Republic. The British started to apply the name 'dollars' to several other currencies, including those of Portugal and Spain, and it evidently stuck.

Until the creation of the official US dollar, American citizens were widely using what was known as the 'Spanish dollar'. Even after the newly independent United States started minting its own silver dollars in 1794, the pieces of eight remained in demand. Indeed, a shortage of silver forced the government in 1797 to recognise officially the ongoing validity of the Spanish money.

Pieces of $

The dollar sign – $ – is also said to owe its origins to the *peso de ocho reales*, although this is not 100 per cent certain. The theory goes that when written down, 'Pieces of Eight' was abbreviated to 'P8' or '/8/'.

What happened next was supposedly that either the stalk of the 'P' from the first abbreviation, or the oblique strokes of the second, were overlaid on the '8' as a further shortcut. As a nod to the legacy of pieces of eight being cut up for small change, the '8' then became 'S' – not the letter S, but rather a symbol showing an '8' with bits missing.

It should be stressed that this is one of a number of theories on the origin of the $, and anyone who tries to persuade you they know the definitive answer should be made to walk the plank.

> **Pieces of eight are still highly lucrative today. Coins salvaged from the wreck of the *Santa Maria de la Consolación*, which sank near the Isla de Muerto (Island of the Dead) in the Bay of Guayaquil, Ecuador, in 1681, often fetch prices of up to $1,000.**

180
Oooone Hundred and Eiiiiightyyyyy!

D arts players are often impressive for two key reasons: firstly, the best players seem to have somehow figured out the ratio of the perfect-sized beer belly to the most accurate throw. Secondly, they have an ability to do impressive feats of maths in the blink of an eye to decide which part of the board

they are going to aim towards next. (For those not educated in the ways of 'arrows', the 180 to which we are referring is the top score you can get with three darts.)

Scraping the Bottom of the Barrel

The origin of darts is unclear but it may well have evolved hundreds of years ago with English archers looking to find new ways to improve their skills. ('Tristan, thou can'st fire thine arrows true, but can'st thou challenge me in chucking them at thon barrel yonder? Come and have a go, if thou think'st thou art hard enough.')

There are tales of the original boards being made of the bottom of barrels or offcuts from felled trees. Both theories give rise to the idea that the rings on a dartboard represent the rings found in wood.

The dartboard as we know it wasn't created until the 1930s, when the 'bristle board', made of sisal fibre, began to take the place of its wooden predecessor. It was attractive not least because old wooden boards needed to be soaked in water overnight to stop them cracking. But the new board didn't become popular until the 1970s because it was considerably more expensive.

Perfect Randomness

It is said that the numbering system on the modern dartboard was created in 1896 by an English carpenter called Brian Gamlin. Not much is known about him, something darts historians (yes, they do exist!) say could be down to the fact he was part of a travelling circus, so was constantly on the move.

Gamlin's layout was designed to minimise 'lucky throws' and promote skill by placing low numbers next to high numbers. Hence number 20 is flanked by the numbers five and one.

It's not clear how long he took to settle on his design but it might have been a while since, as darts historian Patrick Chaplin (told you they existed!) explains, there are 2,432,902,008,176,640,000 different possible arrangements of the segments on a standard dartboard.

Despite Gamlin's humble origins, darts aficionados say the layout is close to 'perfect randomness', a phrase well suited to describe the direction in which many an amateur darts player finds their darts travelling through the air.

A Load of Bull: Darts in Numbers

- The dartboard should always measure 5 feet 8 inches from the centre of the bullseye to the ground (that's 68 inches or 173 centimetres).
- The distance from the front of the dartboard to the throwing line (the 'oche') should be 7 feet 9 and a quarter inches (93.25 inches or 237 centimetres).
- The target of 501 evolved from a time when scores were measured using a cribbage board. 'Five times round the cribbage board', as the score once went, was equivalent to 301. This later evolved into the longer game of 501 once chalkboards were introduced, according to Patrick Chaplin.
- 'Three in a bed' is slang for getting all three darts into one number.
- The left-hand side of the board is known as the 'married man's side' because it has more high numbers and is therefore less risky.

*#06#

THE KEY TO YOUR TELEPHONE'S SECRET NUMBER

No doubt you can remember all the digits in your mobile phone number – and probably in the right order too. But there's another number you should have memorised, or at the very least written down somewhere safe, and that is your phone's IMEI number.

This stands for International Mobile Equipment Identity number and it is a great way of getting the last laugh if someone steals your phone. If this does happen, all you need to do is call your service provider and tell them to block your phone using the IMEI number. This renders the phone useless and means the nasty criminal will be unable to run up a big bill at your expense.

There are three straightforward places to look to find out what your IMEI number is:

1. Under the battery.
2. On the side of the box the phone comes in.
3. Dial *#06# into your phone and the IMEI will appear on the screen.

We might not be able to stop criminals stealing handsets, but we can certainly try to annoy them as much as possible when they do.

66

GET YOUR KICKS
ON ROUTE 66

There are few more iconic places you could go for a drive than along Route 66, the 2,448-mile highway that ran between Chicago and Los Angeles before it was officially decommissioned in 1985.

This road holds a special place in the hearts of Americans, as well as many who have never been anywhere near it. It symbolises pioneering spirit, freedom, opportunity and adventure in a way no other road probably ever will again.

Paving the Way to Freedom

US Route 66 was commissioned on 11 November 1926. If it had not been for various political wrangles it would have ended up being called Route 60 or Route 62.

It took over ten years for the road to get paved from end to end, something that happened in 1937. This might sound late but you have to remember it was only a few decades previously that motorcars first came into being and people had to walk in front of them to make sure they didn't scare oncoming horses.

The Mother of All Roads

This road, which spans three time-zones, became a source of hope in the 1930s for hundreds of thousands of people escaping the 'Dust Bowl', a series of turbulent dust storms that had rendered millions of acres of farmland useless.

This exodus was immortalised by John Steinbeck in *The Grapes of Wrath*. He wrote: '. . . and they come into 66 from the tributary side roads, from the wagon tracks and the rutted country roads, 66 is the mother road, the road of flight.'

It was after this the highway became known as the 'Mother Road', but it was also regularly known as Main Street of America, and the Will Rogers Highway (after the 1930s film star).

> Numerous other popular culture references helped immortalise the road, including Bobby Troup's song '(Get Your Kicks on) Route 66', and the eponymous 1960s TV show. In *On the Road*, Jack Kerouac famously observed the 'long-haired broken-down hipsters straight off Route 66 from New York'.

You Can Still Get Your Kicks (Just About)

As time passed, the growing demand that traffic was putting on the road system led to upgrades on the route. But it also led to the building of multi-lane highways, eventually rendering Route 66 largely obsolete in the eyes of the travelling public and the authorities.

It was decommissioned in 1985 and much of the road, with its 'Ma and Pa' shops and 'full-service gas stations' fell into disrepair. However, recently a number of societies have sprung up dedicated to protecting and improving Route 66.

These days it is possible to travel on around 85 per cent of the old road, which means adventurers can still enjoy the sense of freedom and the thrill of seeing the iconic road signs that line the route.

.44 MAGNUM
DO YOU FEEL LUCKY?

There are few more memorable lines in silver screen history than these ones, penned by screenwriters Harry Julian Fink, R M Fink and Dean Riesner, and delivered by Clint Eastwood in the 1971 film *Dirty Harry*, as he brandishes his .44 Magnum gun at a hapless robber, having just shot and killed his two accomplices:

> I know what you're thinking. 'Did he fire six shots or only five?' Well, to tell you the truth, in all this excitement I kind of lost track myself. But being as this is a .44 Magnum, the most powerful handgun in the world, and would blow your head clean off, you've got to ask yourself one question: 'Do I feel lucky?' Well, do ya, punk?

As it happens, after the robber has surrendered, Dirty Harry reveals the gun chamber was empty – not that that would have stopped Clint doing this villain some very serious damage if he'd wanted to. Anyone who has ever held this gun will know that just hitting the bad guy on the head with it would cause major injury: the weapon is very, very heavy.

Bullet Points

The gun's full name was Smith & Wesson Model 29 .44-calibre Magnum revolver. The '44' part indicates the diameter of the bullet that the gun fires, measured in hundreds (or sometimes thousandths) of an inch. Due to the need for a snug fit, the inside measurement of the barrel usually matches this.

So, to use another example, a .25 calibre bullet is a quarter of an inch in diameter. Some calibres, particularly on European-made guns, are measured in millimetres, giving Arnold Schwarzenegger the opportunity to memorably order an 'Uzi 9 mm' in *Terminator*.

Explosive Power

But as Arnie no doubt knows, size isn't everything. Just because a round has a larger diameter, it does not necessarily follow that it is more powerful. Power depends not only on calibre but also other factors such as muzzle velocity and the amount of explosive in the cartridge.

This is reflected in the 'Magnum' element of the name, which denotes that the gun fires a particularly powerful round of ammunition (other manufacturers use the word 'Special' instead).

How to Kill Someone Even Deader

Dirty Harry's claim that the .44 Magnum was the 'most powerful handgun in the world' stood up at the time but it has since been eclipsed.

To give you an idea of how things have moved on, experts say that in 1971 Harry's .44 Magnum would have produced about 900 foot pounds of muzzle energy. (Foot pounds – ft-lb – are a standard way of measuring the power of revolvers.) In comparison, Smith & Wesson claim their 500 S&W Magnum, launched in 2003, is now the world's most powerful handgun. Its 50-calibre bullets are almost two inches long and when fired it produces almost 2,600 ft-lb – almost three times that of Dirty Harry's piece.

That wouldn't just make Harry's day, but his week and month as well, you'd imagine. Just try not to think what it would do to the poor 'punk' staring down the business end of the barrel.

21/12/2012
THE END OF THE WORLD

At the time of writing, the editors and author of this book very much hope that you get it as a Christmas present at the end of 2012. This means the undercurrent of sarcasm and cynicism you are about to encounter while reading about the end of the world has been justified.

If the world did indeed end just before Christmas then it won't matter, and no, you can't have your money back.

Time Is Running Out

The reason 21 December 2012 has been mapped out to be a disaster of biblical proportions by doom-mongers is due to the cycles of the Mesoamerican calendar, as used by the Mayans.

> **The Mayan civilisation existed from around 250–900 AD in what is now Central America. Their traditional calendar spanned 52 years, roughly the length of a person's life. To put greater lengths of time into context they also created something called the 'Long Count Calendar'.**

The Mayans set their 'Long Count Calendar' to begin in the year 3114 BC (if you apply our modern Gregorian calendar to it). This date was when the Mayans believed the world was created in its current form.

There is general agreement that this Long Count Calendar will 'run out' after 5,126 years. Do you see what's coming? Of course you do. Now run for the hills – if there are any left – or just relax and read on.

If the Long Count began in 3114 BC and it goes on for 5,126 years, the 'end date' would fall in 2012 AD, and more particularly, on 21 December, the date of the winter solstice.

The End Is Nigh!

When these apocalyptic theories surfaced, scholars were quick to point out that the Mayans never said this was the end of the world, merely the end of this particular calendar. As such, in the same way you get a new calendar full of cute kittens each year from a well-meaning relative, so the Mayan Long Count would simply be ending one cycle, to begin a fresh one.

But such sensible analysis should never be allowed to stand in the way of a good 'End of Days' conspiracy theory (particularly if you're trying to sell a book, Hollywood blockbuster, or just set up a successful doomsday cult). Some of the top concerns include the following (and extra exclamation marks have been added to get into the hysterical spirit of things):

- We're going to be cremated by giant flares from the sun!!
- The Earth's magnetic field will reverse!!
- Saturn and Jupiter will align with the Earth causing massive meteorological upset!!
- We're going to get sucked into a black hole in the Milky Way!!
- A giant comet is coming to destroy life as we know it, with the most popular candidate being the 900-foot-wide 'Apophis'!!
- A rogue plant called Nibiru is going to hurtle into us – we've seen the pictures of it approaching and we're pretty sure they're not fake!!

In a nutshell, scientists have answered these outbursts in the following ways:

- A: This may well happen but it has happened before (often lots of times and on a greater scale than you're predicting) and we're still here;
- B: This simply isn't going to happen. Get off the Internet and get a life.

Of course if it does happen then it will wreck Christmas. But look on the bright side: you won't be forced to look at a calendar full of kittens for another twelve months.

ACKNOWLEDGEMENTS

I have to thank my editor at Virgin Books, Kate Moore, for whom this all started over a pint with 562 stamped on it. Without her enthusiasm, this book would simply not exist. The same goes for Duncan Moore, who was the original inspiration for the project – a manly handshake and an album of absurd 1980s rock music are coming your way.

Praise is also due to Lindsay Davies, who did a fantastic job on the manuscript and seemed to endure my loose grip on basic grammar with great stoicism.

Further thanks go to my wife, Susanna, who offered lots of ideas and has an eye for detail that I could never hope to attain.

Lots of people came up with interesting numbers; from my Mum and Dad, to friends with whom I was trapped hiding from a Siberian weather front. You are too numerous to name – which is somewhat ironic in a book all about numbers. However, all your ideas were very much appreciated.

Finally, thanks to those who made my work much easier in the course of researching this book. This includes Mssrs Duckworth and Lewis of cricketing fame, the people at GS1 UK, FIFA press office, the man in Zimbabwe who let me hold his .44 Magnum and, of course, Bury Metropolitan Borough Council: may your pint glasses always grace our bars.

Other Sources

In the course of writing this book, I picked up facts and figures from all over the place, and then cross-referenced them with lots of other sources. I cannot possibly fit them all in here. However, my chief sources included the following, in no particular order:

The British Museum

The Underworld Speaks, Albin Jay Pollock (1935)

Blackstone's Commentaries, 15th Ed. (1809)

The Metropolitan Police

The work of Professor Schindler, Rutgers Business School

The Secret of Chanel No. 5: The Biography of a Scent, Tilar J. Mazzeo (2011)

The International Organization for Standardization

BBC News online

PatrickChaplin.com

number10.gov.uk

The International Playing-Card Society

The All England Lawn Tennis Club

The International Tennis Hall of Fame

espncricinfo.com

Fédération Internationale de Football Association (FIFA)

The Automobile Association

Forbes Travel Guide

The Hotelstars Union

The International Cloud Atlas (1896)

Dictionary of American Slang (1960)

www.factoryrecords.net

Book of Revelation, The Bible

catholic.net

thatreligiousstudieswebsite.com

'Patristic Evidence of the Use of Nero's Name in Calculating the

Number of the Beast (Rev 13:18)', *Westminster Seminary Journal* 68 (2006)
The World Gold Council
555-LIST
The Scotch Whisky Association
The Royal Naval Museum
London Stock Exchange
urbanlegends.about.com
National Geographic
BARB (Broadcasters' Audience Research Board)
Nielsen
aerospaceweb.org
Golfball-guide.de
Guinness Book of World Records
New York Times
The Skin Cancer Foundation
International Federation of Football History & Statistics
Historicalkits.co.uk
Footballshirtculture.com
1984, George Orwell (1949)
Orwell: Wintry Conscience of a Generation, Jeffrey Meyers (2001)
ukfoodguide.net
exploreenumbers.co.uk
'Seed size variability: from carob to carats', Lindsay A Turnbull, The Royal Society, 2006
The US Secret Service (online)
419eater.com
Master Stamp List, Department for Business, Innovation and Skills
legislation.gov.uk
gunatics.com
popularmechanics.com
National Historic Route 66 Federation
'Ed Pegg Jr.'s Math Games: Sudoku Variations', The Mathematical Association of America
sudokuessentials.com
International Telecommunication Union (formerly CITT)
NASA
The National Archives
British Medical Journal
pottermore.com
Ofcom
forbes.com
drdossey.com
Morris Dictionary of Word and Phrase Origins, William and Mary Morris (1997)
napoleon.org
Dirty Harry (1971)
The Grapes of Wrath, John Steinbeck (1939)